PROCEEDINGS OF THE 30TH INTERNATIONAL GEOLOGICAL CONGRESS
VOLUME 7

OROGENIC BELTS
GEOLOGICAL MAPPING

Proceedings of the 30th International Geological Congress

Proceedings of the
30th International Geological Congress

Beijing, China, 4 - 14 August 1996

VOLUME 7

Orogenic Belts Geological Mapping

Editors:
Xu Zhiqin
Institute of Geology, CAGS, Beijing, China
Ren Yufeng
Institute of Geology, CAGS, Beijing, China
Qiu Xiaoping
Institute of Geology, CAGS, Beijing, China

///VSP///

Utrecht, The Netherlands, 1997

VSP BV
P.O. Box 346
3700 AH Zeist
The Netherlands

First published in 1997

ISBN 90-6764-263-0

Printed in The Netherlands by Ridderprint bv, Ridderkerk.

CONTENTS

PREFACE

The orogenic belt which is described as "Natural Laboratory" of continental dynamics, is an important studying domain in Earth sciences. Since the 1960s, the Plate Tectonic Theory, named "Geoscience Revolution" has brought about a new intension to the orogenic belt, i.e. the orogenic belt is the result of a collision between lithosphere plates. In recent years, lots of researches have shown that the huge deformation and reorogeny can occur within a continent, and the uplift and exhumation of post-orogeny has been considered as an important content. With the study of deep geophysical probes and deep physical processes, it is concealed that orogenic dynamic mechanism can not be simply interpreted by the classical Plate Tectonic Theory and Model. The new stage of finding a new driving force within continents and the exploring genesis of orogenic belt has started.

The geological map is the comprehensive expression of studying results of regional geological surveys, and is the important means for geological study. With the applications of computer science and technology, geological mapping is becoming more and more efficient and facile. This, on the contrary, will greatly impel the development of geoscience.

Beijing
June 1997

Xu Zhiqin

Proc. 30th Int'l Geol. Congr., Vol.7, pp. 1-6
Xu Zhiqin, Ren Yufeng and Qiu Xiaoping (Eds.)

Secondary Compaction through Area Loss during Accretion; A Sand Model Approach

HEMIN KOYI
The Hans Ramberg Tectonic Laboratory, Institute of Earth Sciences, Uppsala University, Uppsala, Sweden.

Abstract

Loss of area (volume) through secondary (tectonic) compaction is quantified by analysing sequential sections of a sand model shortened above a rigid substratum. Area balancing shows that the model wedge tectonically compacts during deformation and it looses up to 17% of its area at 50% shortening. During the deformation, the wedge does not undergo homogeneous compaction. Instead, the incremental compaction increases towards foreland (toe of the wedge) whereas the accumulative compaction increases towards the hinterland (rear of the wedge) most of the area loss takes place within the youngest of the imbricates. Between two stages of imbricate formation, the model accretionary wedge lost 3.9% of its area of which 3% was taken up within the youngest of the imbricate sheets at the toe.

Length balancing of three layers located at different depths (top, middle and bottom) were conducted to determine and compare the three strain components (layer parallel thickening, folding and imbrication) at different stratigraphic levels in the model accretionary wedge. The shallow top layer has taken 37% shortening by imbrication and folding and only 17% by layer parallel thickening as compared to 13.6% shortening by folding and imbrication and 41% shortening by layer parallel thickening for the deeper bottom layer. The middle layer has intermediate values. These results suggest that ductile deformation dominates at depth. In natural accretionary wedges, area loss and layer-parallel thickening is accommodated for by secondary (tectonic) reduction of porosity, pressure solution of minerals and development of fabric.

Keywords: secondary compaction, area loss, sand model, different bulk shortening

INTRODUCTION

Many workers have used sand models to study different aspects of accretionary wedges [5, 3, 7, 13, 11, 1, 10, 9, 6]. Davis et al. [5] and Dahlen et al. [2] presented a comprehensive study on the dynamics of accretionary wedges and compared them to deformation of sand or snow in front of a bulldozer. In an earlier paper [9], different aspects of internal deformation within a model accretionary wedge were discussed. This follow up article illustrates the significance of area loss and penetrative strain within a sand-model accretionary prism. It is shown here that not only area balancing does not restore a model profile to its initial stage, but also that length balancing of two layers located at different depths restores them to two different initial lengths, which gives variable amount of bulk shortening.

MODEL DESCRIPTION AND DEFORMATION

A sand model consisting of passively layered loose sand was shortened by 47 % from one end (Fig. 1). The model contained passively coloured homogeneous sand with a constant thickness of 5 mm [12]. Profiles were sequentially eroded by the means of a vacuum cleaner throughout the deformation at every 1.5% shortening for photographing. This technique exposed the three-dimensional geometry of the imbricate sheets and allowed using small grain size sand and sand layers as thin as 0.2 mm thick to quantify strain in detail. Sequential sections were prepared and photographed during the deformation of the model. These sections are used to quantify area loss and penetrative strain in the model. Loose sand, which has a low friction and show Coulomb behaviour is an appropriate analogue material to simulate deformation in the rocks of the upper brittle crust. This study is a two-dimensional approach assuming no movement in the (Y-axis along strike) because movement perpendicular to transport direction is small relative to the movement parallel to it. In the experimental approach I have deliberately left out the additional complicating factors of erosion, deposition, material anisotropy due to facies changes, time variations in pore fluid pressure ratios across the wedge and changes in friction coupling along the decollement surface. The model described here, was deformed on a horizontal planer and rigid substrate and therefore does not account for natural decollement slope or irregularities. For more details of model deformation and sectioning, see [9].

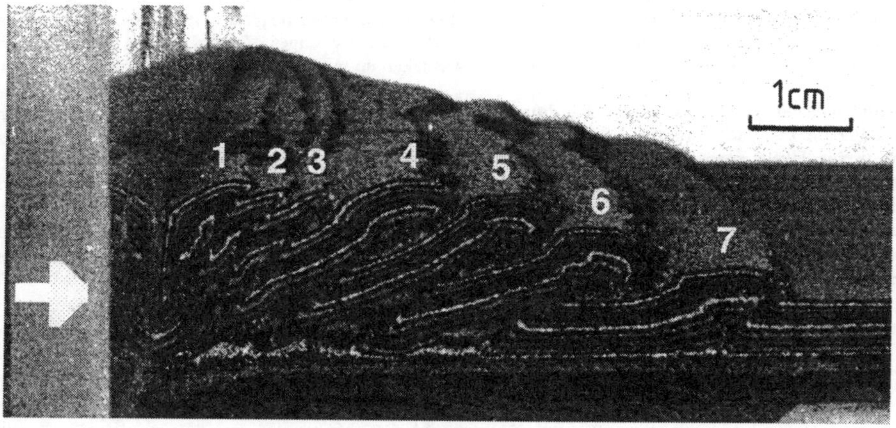

Figure 1. A photograph of the sand model after 40% shortening showing seven imbricate sheets. The numbers indicate the sequence at which the imbricates have formed. Arrow shows direction of shortening.

AREA LOSS AND STRAIN PARTITIONING

Procedure

In a profile of the final stage of the deformed model, bed-length balancing was conducted for three layers located at different stratigraphic horizons (Fig. 2). Bed-length balancing was used to partition the amount of deformation taken up by the three strain components; layer parallel thickening, kink folding and imbrication. This was prepared by measuring the length of segments of each deformed layer and adding them together.

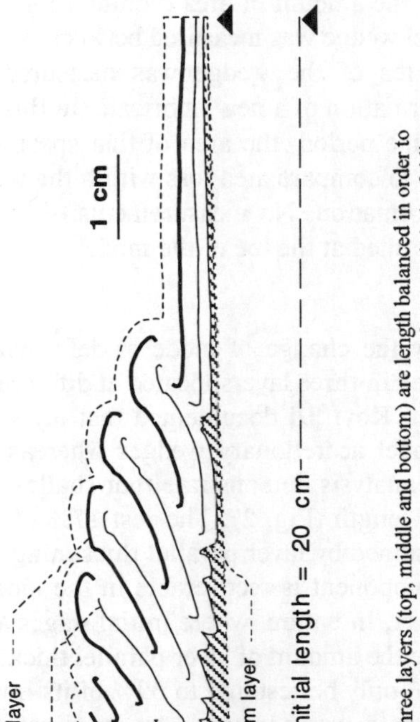

Figure 2. A line drawing of model profile at 44 % bulk shortening. Three layers (top, middle and bottom) are length balanced in order to quantify the three components of strain (layer parallel shortening, folding and imbrication). Note that the top layer deforms mainly by folding and imbrication in contrast to the bottom layer which deforms mainly by layer parallel thickening

Consequently, the effects of imbrication and folding were taken out by joining the displaced segments of the layers together and unfolding them (Fig. 2). By comparing this "restored" length with the deformed length, the amount of shortening by imbrication and folding was calculated (Fig. 2). Then, this "restored" length was compared to the initial (undeformed) length of the model to calculate the amount of shortening taken up by layer parallel thickening.

In order to quantify the amount of area (volume) loss in the model during deformation, the area of the model wedge was measured between two stages of imbricate formation. In other words, the area of the wedge was measured at the onset of formation and throughout the deformation of a new imbricate (in this case imbricate number 4, Fig. 2). During the same time period, the area of that specific imbricate (number 4) was also measured with time to compare area loss within the wedge and the imbricate during the same period of deformation. No measurements of area loss were included as soon as another imbricate formed at the toe of the model.

Discussion
In order to quantify the change of mode of deformation with depth, the longitudinal strain was partitioned in three layers located at different stratigraphic levels (top, middle and bottom). Earlier, Koyi [9] documented that layer parallel thickening dominates in deeper levels of model accretionary wedges whereas imbrication prevail at shallower levels. The current analysis demonstrate that shallow layers could be restored back to 83% of their initial length (Fig. 2). The rest 17% of the initial length, which was not restorable, was consumed by layer parallel thickening during deformation of the model. This latter strain component is recoverable in the model because its initial dimensions were known. However, in nature, where initial stages are not well documented, it would be difficult to restore the amount of layer parallel thickening which increases with depth. The deep layer could only be restored to 59% of its initial length. The entire unrestored 41% of the layer length was consumed by layer parallel thickening. This comparison does not only illustrate that the mode of deformation within an accretionary wedge changes with depth, but it also points out that bed length balancing of different layers in natural profiles of accretionary wedges may give different percentage of shortening. As deeper layers could be restored to shorter initial lengths than shallower layers, they show more percentage bulk shortening. Therefore, to determine the bulk shortening of natural examples, bed length balancing should be exercised on layers located at different stratigraphic horizons.

At the onset of model deformation, the bulk shortening is accommodated by penetrative strain, which result in thickening of the undeformed sand layers. This stage is followed by kinking of these layers as shortening continues. Mulugeta and Koyi [12] demonstrated that volume loss in a model accretionary wedge is episodic corresponding to the episodic growth of the wedge. Here, it is shown that at any given stage of model deformation, area (volume) loss is not constant, but varies both laterally and with depth. In the earlier section, the depth variation of strain, which indirectly illustrates change in area loss with depth, was explained. In this section the lateral variation of area loss within the wedge is documented. As it is suggested for Nankai accretionary prisms [8], where shortening of the soft sediments in the prism toe is pervasive and is not only

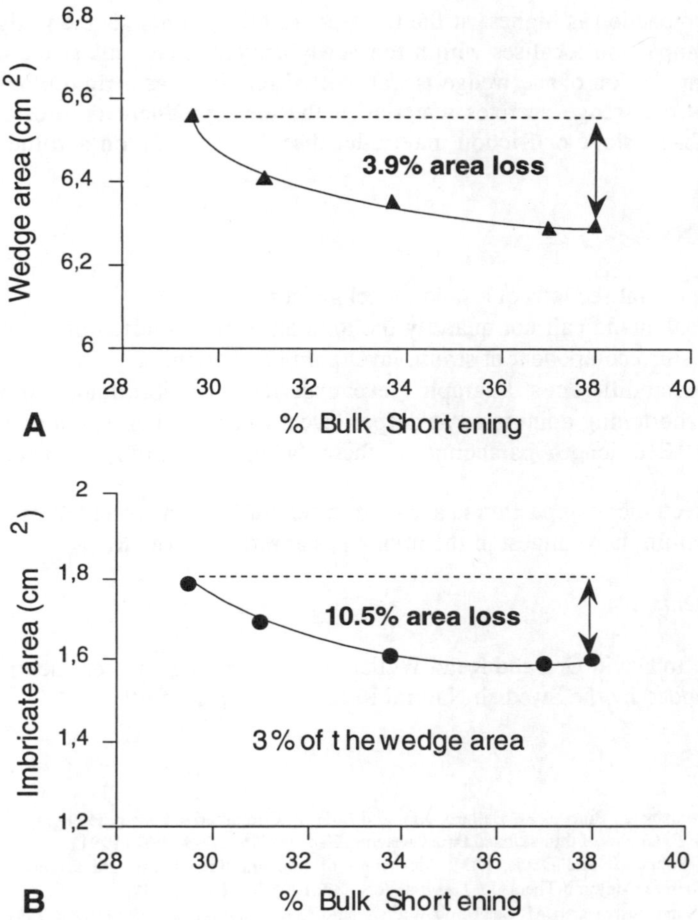

Fig. 3. Plots of (A) wedge area and (B) imbricate area with bulk shortening. imbricates. At 38% shortening, the wedge loses ca. 4% of its initial area. During the same period, the youngest imbricate loses 10.5% of its initial length which is equal to ca. 3% of the total area of the wedge. This suggest that much of the area loss within the wedge is taken up within the youngest of the imbricates at the toe.

accommodated by displacement along imbricate thrusts and shear bands, the active toe area in the model wedge stands for most of the penetrative strain which the wedge experiences. Area balancing of the model wedge between two stages of imbricate formation show that 76% of the total area loss within the wedge was accommodated within the youngest of the imbricates at the toe (Fig. 3). During the same period of deformation, the entire wedge loses only 3.9% of its area of which 3% was taken up by the youngest of the imbricates. This imbricate sheet loses 10.5% of its initial area during the same period of deformation (Fig. 3). These results suggest that the newly accreted materials at the toe of the wedge undergo secondary tectonic compaction whereas the rear part of the wedge, which contain older stack of imbricate sheets, are strongly compacted and can not accommodate much more tectonic compaction with deformation. In general, cumulative compaction is highest at the rear of the wedge whereas incremental compaction is highest at the toe area. In other words, as the wedge grows, incremental compaction localises within the newly accreted materials at the toe of the model. The rear portion of the wedge is transported relatively as a rigid block without much strain as the wedge accretes material at the toe. Synkinematic erosion or any change in the basal slope or friction may re-localise the deformation and thereby alter this scenario.

CONCLUSIONS

Analysis of sequential sections of a sand model suggest that:
1. Bed-length balancing can not quantify the total amount of shortening as it does not account for the third component of strain, layer parallel shortening.
2. Layers located at different stratigraphic level experience variable amount of strain. Layer parallel shortening dominated at deeper levels whereas imbrication dominates at shallow levels. Bed length balancing of these layers give different percentage of shortening.
3. Secondary (tectonic) compaction is accommodated for by area (volume) loss, which is concentrated within the youngest of the imbricates at the toe of the wedge.

Acknowledgements

Thanks are due to David Gee and Ruud Weijermars for reading and commenting on the manuscript. Funded by the Swedish Natural Research Council (NFR).

REFERENCES

1. Colletta, B., Letouzey, J., Pinedo, R., Ballard, J.F., and Bale, P. Computerized X-ray tomography analysis of sandbox models: Examples of thin-skinned thrust systems, *Geology,* **19,** 1063-1067 (1991).
2 Dahlen, F. A., Suppe, J. and Davis, D:M. Mechanics of fold and-thrust belts and accretionary wedges (continued): Cohesive Coulomb Theory, *J. Geophy. Res.*, **89,** 10087-10101 (1984).
3. Dahlen, F. A. Noncohesive critical Coulomb wedges: and exact solution, *J. Geophy. Res.*, **89,** 10,125-33 (1984).
4. Dahlstrom, C.D.A. Balanced cross sections. *Can. J. Earth. Sci.*, **6,** 743-747 (1969).
5. Davis, D.,Suppe,J. & Dahlen, F.A. Mechanics of fold-and-thrust belts and accretionary wedges. *Jour.Geophy. Res.*, **88,** 1153-1172 (1983).
6. Gatscher, M.A., Kukowiski, N., Malavieille, J. and Lallemand, S. Cyclical behaviour of thrust wedges: Insight from basal friction sandbox experiments. *Geology*, **24,** 135-138 (1996).
7. Karig, D.E. Physical properties and mechanical stateof accreted sediments in the Nankai Trough, Southwest Japan Arc. *Geol. Soc. Am. Memoir*, **166,** 117-133 (1986)
8. Karig, D.E. and Lundberg, N. Deformation bands from the toe of the Nankai Accretionary prism. *J. Geophy. Res.*, **95 (B6)**, 9099-9109 (1990).
9. Koyi, H. Mode of internal deformation in sand wedges, *Journal of Structural Geology*, **17,** 293-300 (1995).
10. Liu, H., McClay, K.R. and Powell, D. Physical models of thrust wedges. In: *Thrust Tectonics* (edited by McClay, K.R.), 71-81 (1991).
11. Mulugeta, G & Koyi, H. Three-dimensional geometry and kinematics of experimental piggyback thrusting, *Geology*, **15,** 1052-1056 (1987).
12. Mulugeta, G & Koyi, H. Episodic accretion and strain partitioning in a model sand wedge, *Tectonophysics*, **202,** 319-333 (1992).
13. Zhao, W. L., Davis D.M., Dahlen, F.A. & Suppe, J. Origin of convex accretionary wedges: evidence from Barbados, *J. Geophys. Res.* **91,** 10246-10258 (1986).

Proc. 30th Int'l Geol. Congr., Vol.7, pp. 7-20
Xu Zhiqin, Ren Yufeng and Qiu Xiaoping (Eds.)

Mesozoic Crustal Evolution and Dynamics of the East Kunlun — Tanggula Composite Mountain Chains

XU ZHIQIN YANG JINGSUI LI HAIBING XU QIANG CHEN WEN GUO ZHENGFU
Institute of Geology,Chinese Academy of Geological Sciences, Beijing 100037,P.R.China

Abstract

The East Kunlun-Tanggula Mesozoic composite mountain chains are composed of the kunlun terrain (North Kunlun terrain and South Kunlun terrain), Bayan Har-Songpan-Garze terrain and the Qiangtang-Tang gula terrain. The Successive subduction and consumption of the Paleotethys and Neotethys led to the comergence and collision of the Kunlun, Bayan Har-Songpan-Garze, Tanggula-Qiangtang terrains and Gondwana and resulted in the Mesozoic orogenic process, forming the Paleotethyan and Neotethyan Composite imbrication mountain chain.
The Paleotethys (the Anymaqen and Jinshajiang Oceans) opened in the Paleozoic-Early Triassic, separating the kunlun, Bayan Har-Songpan-Garze and Tanggula-Qiangtang terrains. The Paleotethys Subducted during the Early to Late Triassic. The Anymaqen oean subducted northward under the Kunlun terrain, the Jinshajiang ocean subducted southward under the Qiangtang-Tanggula terrain, due to the difference of ocean subducted angle, the subduction system of differential tectonic environment are formed. The three terrains collidel at the and of the Triassic, forming the 500km - wide northern lndosinian orogenic belt. The Bangong-Nujing orean opened in the Jurassic, separating the Tanggula-Qiangtang terrain and Gondwana. The Tanggula-Qiangtangterrain collided with Gonduana at the and of the Jurassic, forming the southern Tanggula orogenic belt. The East Kunlun-Tanggula composite mountain chains formed. Since the Cretaceous, the Eart Kunlun-Tanggula region is in uplifting.

Key words: Tethy Oceanic Basin, ophiolite, Subduction System , Orogeuic Process, Crustal Structure, Mesozoic, East Kunlun-Tanggula Composite Mountain Chains.

INTRODUCTION

Since the Late Paleozoic — Early Triassic, there occurred between Eurasia and Gondwana a series of important geologic events such as the opening of the Tethys oceanic basin and the splitting of continent, subduction, consumption and closing of oceanic crust, collision and convergence between continents, orogenic amalgamation, crustal shortening and thickening and the rotation of blocks. All these events have been recorded in the East Kunlun — Tanggula Mesozoic composite mountain chains, which provides the key evidences and geologic features to study the Mesozoic crustal evolution and dynamics.

TETHYS OCEANIC BASIN AND OPHIOLITE

The East Kunlun — Tanggula Mesozoic composite mountain chains are composed of the

North Kunlun terrain, South Kunlun terrain, Bayan Har — Songpan-Garze terrain and the Qiangtang terrain. The Central suture zone between the North Kunlun and South Kunlun terrains is the product of the closing of the oceanic basin between the North China plate and South China plate and the collision of these two plates before the Late Devonian.

Since the Late Paleozoic, the successive opening of the Paleotethys and the Neotethys resulted in the splitting of Gondwana and southern Eurasia, which is primarily marked by the presence of ophiolite. There are four Tethyan ophiolite belts in the present Qinghai-Tibet plateau, the Paleotethyan Anymaqen belt(P_2 — T_{1-2}), the Paleotethyan Jingshajiang belt (P_2 — T_{1-2}), the Neotethyan Bangong-Nujiang belt (T — J_2) and the Neotethyan Yarlung Zangbo belt (K — E). The former three belts lie within the East Kunlun-Tanggula Mesozoic mountain chains, while the later is located at the boundary between the Gandise-Nyanqentangula terrain and the Indian continental block. The Anymaqen oceanic basin split the Yangtze continent from southern Eurasia and divided it further into the South Kunlun terrain and Bayan Har — Songpang-Garze terrain. This small "shear-type" oceanic basin together with the contemporary Jinshajiang oceanic basin to the south made the Bayan Har terrain as a continent bridge extending into the oceanic basin (Fig. 1).

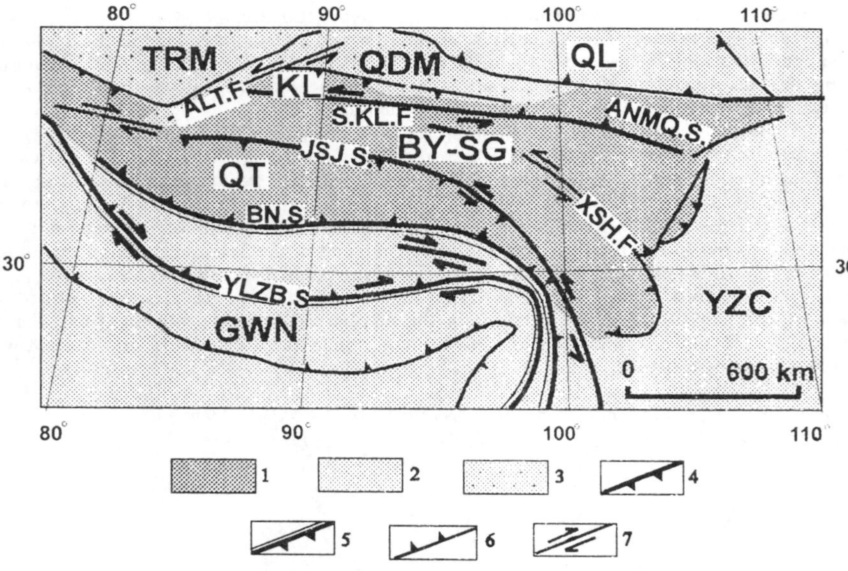

Figure 1. Plate tectonic sketch of the East Kunlun-Tanggula Mesozoic composite mountain chains.1. scope of Mesozic orogenic belts; 2. surrounding terrains; 3. Mesozoic-Cenozoic basins; 4. Paleotethyan sutures; 5. Neotethyan sutures; 6. thrusting faults; 7. strike-slip faults; E. KL — East Kunlun terrain; BY-SG — Bayan Har-Songpan-Garze terrain; QT — Qiangtang-Tanggula terrain; GWN — Gondwanaland; YZC — Yangtze craton; TRM — Tarim basin; QDM — Qaidam basin; QL — Qilian Mountains, S.KL.F — Southern Kunlum fault; XSH.F — Xianshuihe fault; ANMQ. S — Anymaqen suture; JSJ. S — Jinshajiang suture; NB. S — Bangong-Nujiang suture; YLZB. S — Yarlung Zangbo suture

(1) Anymaqen ophiolite belt

The Anymaqen ophiolite belt, located in the southern margin of the Kunlun terrain, extends from Muztagh eastward to Maqin for about 1000km.

Jiang and Yang (1992) named it first and, based on the 260Ma age of the basalt in the Xidawu region of Maqin and the foraminifera in the radiolarian silicalite, determined that it began in the Late Permian. Recent studies [16] show there is mid-oceanic ridge basalt (MORB) all in the Buqinshan, Majixieshan and Maqin in the eastern section of the ophiolite belt with deep-sea Mg-rich metamorphosed periodic, bedded gabbro of cumulate texture, plagiogranite, gabbro and cumulate peridotite. The tholeiitic dyke swarms marking sea-floor spreading found in the Maqin region further demonstrate that the ophiolite suggests an oceanic basin. Because of dextral transformation in the western section, the outcrop of ophiolite has only been found in Muztagh [6]. This belt may extend eastward via Nanping and Kangxiang, and join the West Qinling Mianxian — Lueyang ophiolite belt [19] (Fig.2).

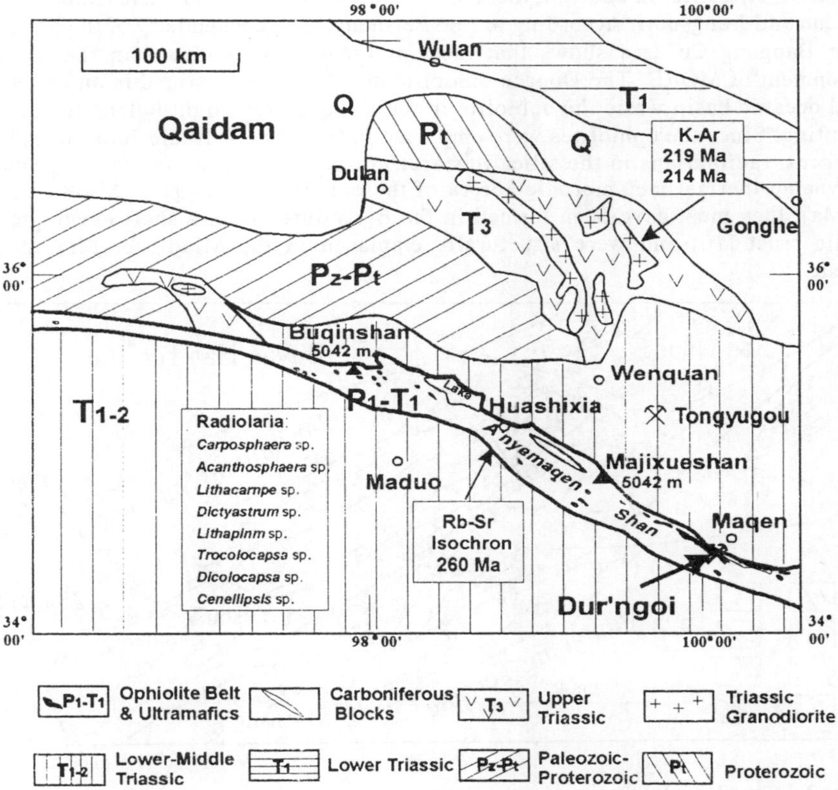

Figure 2. Sketch geological map of the Anymaqen ophiolite belt

(2) Jingshajiang ophiolite belt

The Jingshajiang ophiolite belt is east-west striking between the Bayan Har terrain and the Qiangtang terrain and north-south striking in the western margin of the Indosinian Songpan-Garze orogenic belt. The latter north-south striking section consists of the eastern Garze-Litang and western Ailoshan-Batang belts. The formation age of the ophiolite in the eastern belt is believed to be P_2 — T_2, but that of the western belt is controversial (C — P_1,) [14] though it indisputably belongs to the Paleotethys. No outcrop of typical opiolite has been found in the considerably long western belt and our recent studies on the dismembered lherzolite, gabbro, pillow basalt and diabase swarms show that they are of marginal volcanic arc rather than oceanic crust [16,17]. We believe that Andes-type oceanic crust subduction may have took place in the Yushu region [14,15] (Fig. 3).

(3) Bangong — Nujiang Ophiolite Belt
Starting from the Bangong Co in western Tibet through Gerze, Dongqiao and Denqen, turning southward, to Myanmar along the Nujiang River, this belt contains 80 ophiolite complexes. The ophiolites in the Bangong Co and Denqen are more or less complete, consisting of mantle peridotite, bedded gabbro, massive and pillow basalt and radiolarian silicalite. In addition there are cumulate complex and basic dyke swarms in Dongcuo and Jiangsuori. According to Bao Peisheng, the geochemistry of the basic lavas in the Bangong Co area shows that most of them were produced in the spreading environment of MORB. The Dongco ophiolite may have been formed in an extensional initial oceanic basin while the ophiolite in the Denqen area might belong to the island arc setting. Since the ophiolites were emplaced in the Lower-Middle Jurassic and older sequences, radiolarians in the silicalites are Jurassic of age [20] and the amphibole in the dynamo-thermal metamorphic aurora of the ophiolite gave age of Middle Jurassic (179Ma), they must have been formed in the Early Jurassic in a short development of oceanic crust [10] and were structurally emplaced in the Middle or Middle-Upper Jurassic.

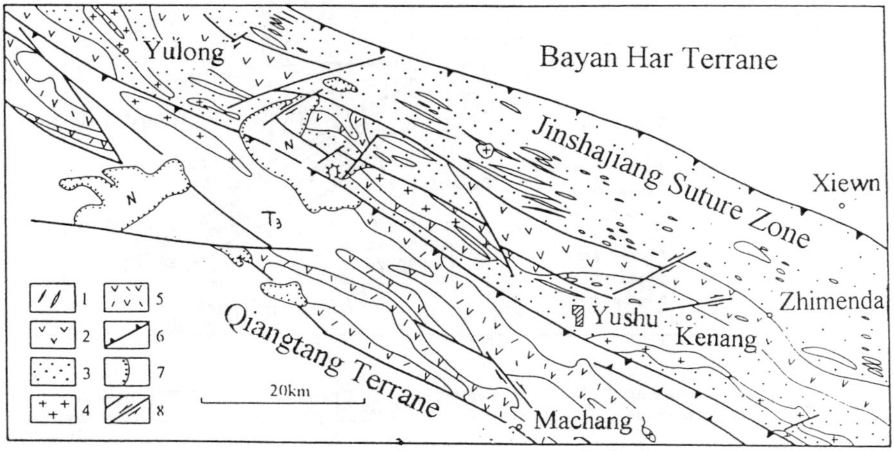

Figure 3. Sketch geoloigical map of the Jinshajiang suture (Yushu region).1 ultrabasic rock and gabbro; 2. epicontinental outer volcanic rock; 3. epicontinental deposit; 4. epicontinental granite; 5. epicontinental inner volcanic rock; 6.thrust fault7.unconformity;8.strike-slip foult and fault.

SUBDUCTION DYNAMICS OF THE TETHYS OCEANIC BASIN

Ocean consumption is realized via plate subduction and the studies on the time-space distribution of the plate regime of the Tethys and kinematics of the subduction rock complexes are indispensable to determining the plate subduction dynamics of the Tethys. Researches suggest that the Anymaqen oceanic crust subducted northward (Triassic), the Jinshajiang southward (Triassic) and the Bangong-Nujiang northward (Jurassic).

Subduction of the Anymaqen Oceanic basin and plate subduction regime of the East Kunlun terrain

A. Volcanic series in the active continental margin of the East Kunlun terrain

On one side of the East Kunlun terrain was the active continental margin resulting from the consumption of the Anymaqen oceanic basin and its northward subduction [3]. Recent studies show that the active margin had a more or less full range of composition and there occurred, from south to north, volcanic rocks formed in the environments of oceanic ridge, island arc, continental margin and collision zone [17], among which the oceanic ridge volcanic rocks occurred in the Buqinshan, Majixueshan and Maqin and the calc-alkaline island arc volcanic rocks (T_1) mainly in the area between Xiadawu and Majixueshan. The island arc had a small area and was immature probably because of the small size of the ocean basin and short-time subduction. North of the island arc there was an enormously thick flysch series of the Early-Middle Triassic age, containing volcanic rocks and olistostrome that represents the sedimentary complex of back-arc basin and continental margin. Among the volcanic rocks those in Tongyugou are basalt formed in back-arc basin and most of the others are the intermediate-acid volcanics rocks of the andesite-dacite-rhyolite assemblage. These volcanics rocks are mainly of the K-rich calc-alkaline series, (Fig.4) rich in Si, poor in Ti, rich in Al_2O_3, poor in CaO and having a lower content of Mg and Fe. REE analysis shows that their $(La/Yb)_N$ is relatively great and Eu anomaly low, suggesting the characteristics of continental margin arc associated with oceanic crust subduction. The contemporary granite belt is mainly of the K-rich calc-alkaline series and the geochemistry of its major elements and its REE and trace element geochemistry all show features of granites in the continental margin arc.

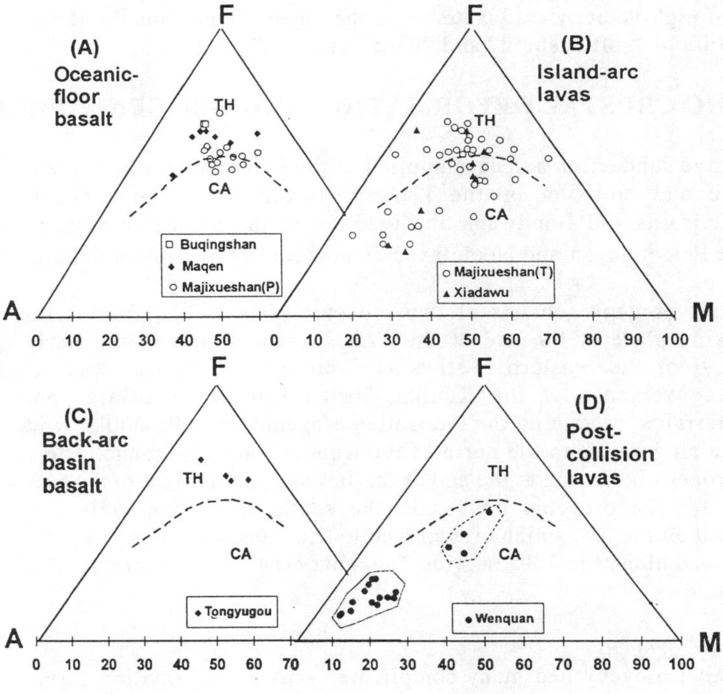

Figure 4. AFM diagram of 4 suites of volcanic rocks in the Anymaqen suture

B. Subduction rock complex
The Anymaqen subduction rock complex is the product of the oceanic crust subduction of the Paleotethys and consists of the Anymaqen ophiolite, melange, fore-arc accretionary prism and high-strain detachment belt. The ophiolite relict composed of ultrabasic rocks, gabbro and tholeiite occur as irregular strips and blocks in the Lower-Middle Triassic sandy slates and which is mixed with exotic blocks of richly fossiliferous Carboniferous and Permian limestones and Triassic Turbidite. The Anymaqen melange belt runs from the Buqingshan Mts. to Maqin on a large scale and with obvious indication.

C. Soft Cushion of Accretionary melange on the passive margin of the Bayan Har-Songpan-Garze Terrain
During the spreading and gradual consumption of the Anymaqen and Jinshajiang oceanic basins (T_1-T^1_3), there formed vast area of sedimentary accumulation mainly of the Triassic flyschoid (the Bayan Har group and the Xikang group) on the Bayan Har continental bridge and the western passive margin of the Yangtze block. The lower part of the Triassic flyschoid series is a typical turbiditic sedimentation with the AB sequence and large olistostrome intercalation. The middle-upper part, mainly of flyschoid, is characterized by the central-outer fan facies and the sedimentary body shows an obvious southward drifting from bottom to top and enormous paleo-current data suggests southward and southeastward flowing. Trace element geochemistry demonstrates that the sedimentary basin was fairly deep and Sengor (1984) called the Triassic sediments in the Bayan Har and Songpan-Garze region as a soft cushion of accretionary melange between the two continents. The data of the Sino-French natural earthquake topographic profile from Golmud to Wenquan show that this terrain is composed of high-velocity cold material in the upper 50km, thus the Bayan Har terrain has a structure of “soft cushion” and “hard basement”.

MESOZOIC CRUSTAL DEFORMATION AND OROGENIC PROCESS

The successive subduction and consumption of the Paleotethys and Neotethys led to the convergence and collision of the Kunlun, Bayan Har-Songpan-Garze, Tanggula-Qiangtang terrains and Gondwana and resulted in the Mesozoic orogenic process , forming the Paleotethyan and Neotethyan composite imbrication mountain chain.

The Mesozoic orogeny consists of two orogenic phases. The first is the Indosinian orogeny taking place at the end of the Triassic, equivalent to the Early Cimmerian orogeny [8,9] of the (eastern) Tethys-Alps orogenic Belt, and was caused by the collisional convergence of the Kunlun, Bayan Har-Songpan-Garze and Tanggula-Qiangtang terrains, producing the Indosinian orogenic belt 500-600km wide associated with Indosinian granites to the north of the Kunlun-Tanggula range. The second is the Tanggula orogeny occurring at the end of the Jurassic, equivalent to the Late Cimmerian orogeny [8,9]. The orogenic belt is in the south of the mountain range and was superimposed on the Indosinian orogenic belt. The Tanggula-Qiangtang terrain collided with Gondwana along the 300Km-wide Tanggula orogenic belt formed of active margin (the north side).

(1) Indosinian orogeny
The Indosinian orogeny had many complicated crustal deformation forms. The major constraints on the forms, intensity and mechanism of the deformation were (1)the competency and thickness of the Caledonian sedimentary cover (Devonian-Triassic) and the paleo-position of the block; (2) the distance from the Paleotethys suture and the

subduction dynamics; (3) the re-activity along the existing structural belts and (4) the direction of the main stress.

(A) Indosinian crustal deformation of the Kunlun terrain
Because of the existence of the Qaidam block during the Early Carboniferous-Late Triassic, there formed around it marginal troughs and bays and thus the cover on the Kunlun terrain was thicker in the east than west and thicker in the south than the north (in the eastern section).

The folding in the western section is simple. In the lower cover (Carboniferous-Permian) there developed syn-cleavage folding with radiating thrust faults and in the upper cover there developed concentric fold. The axis of these folds turned from NW-SE striking in the west gradually to nearly E-W striking in the east, which was associated with the change to sinistral strike-slip of the Paleotethys suture on the southern margin of the Kunlun terrain.

The Indosinian deformation in the eastern section was more complex and there are the following deformation zones from south to north (Fig. 5):

① The frontal subduction thrusting belt of Anymaqen: growing to the north of the Anymaqen subduction zone, it consists of a series of south-verging thrusting faults and nappes. The Indosinian movement (at the end of the Triassic) further reworked the Anymaqen subduction complex, the ophiolite, island arc volcanic rock, sediments in the forearc and backarc basins and the basement of platform facies (Lower Permian limestone) formed before the plate splitting being all involved in the thrusting.

② Backarc detachment north of Maqin: it developed in the flysch series of the Early-Middle Triassic backarc basin north of the Anymaqen subduction zone. The flysch series was slightly metamorphosed to the lower green schist facies and the strong folding and shearing suggest the presence of a southward detachment, in which the flow foliage was further folded into gentle domes.

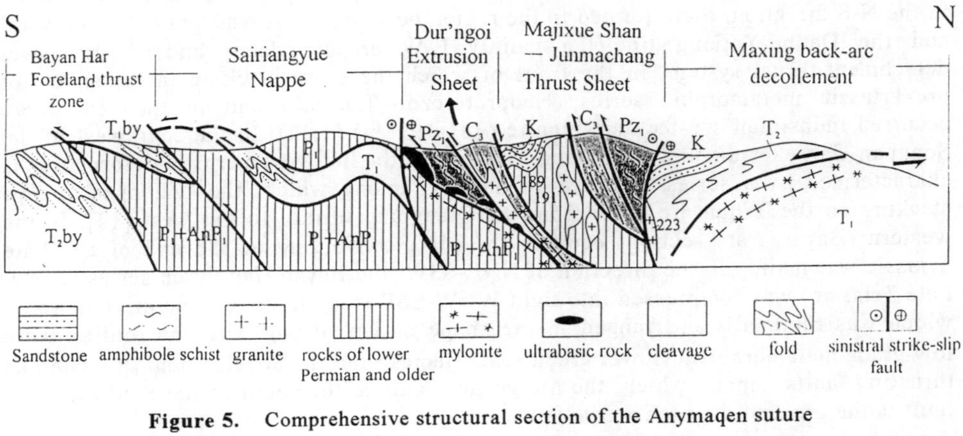

Figure 5. Comprehensive structural section of the Anymaqen suture

③ The central Kunlun thrusting nappe belt: Located in the Lower Paleozoic suture zone of the central Kunlun and to the south, it consists of a series of south-verging thrusting faults and nappes. The thrusting placed the Ordovician-Silurian metamorphic series over the Upper Paleozoic shallow-sea sedimentary rocks which in turn was placed on top of the Lower-Middle Triassic flysch series. There are two large nappes, the Kuhai and Tongyugou, to the south of the central Kunlun belt. Covered by the Jurassic-Cretaceous red beds, the Kuhai nappe is composed of Ordovician-Silurian metamorphic rocks over a ductile detachment under which are the Carboniferous, Permian and Lower-Middle Triassic sequences. This nappe was formed by subsequent intra-continent subduction along the previous structural zones(Fig.6).

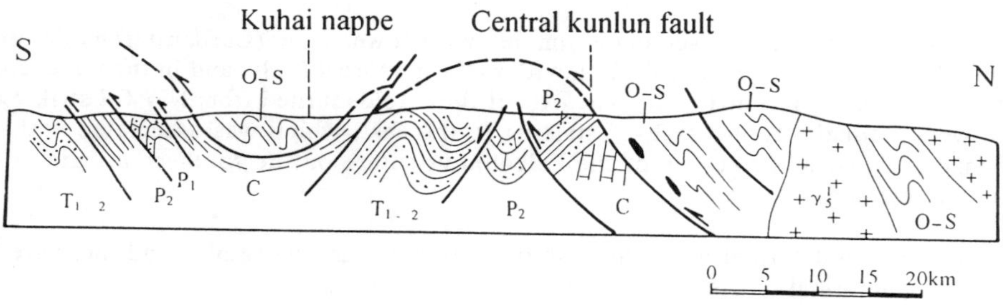

Figure 6. A section of the Kuhai thrust sheet in Wenquan, Central Kunlun Mnt.

④ The northern weak deformation belt: In the vast area from the central Kunlun suture northward to Gonghe, there developed non-foliage vertical concentric folds and reverse faults.

(B) Indosinian Crustal Deformation in the Bayan Har-Songpan-Garze Terrain

During the Indosinian plate converging, the Yangtze continental block subducted as a passive margin separately northward under the Kunlun terrain and southwestward under the Tanggula-Qiangtang terrain, forming the "Bayan Har-Songpan-Garze Terrain" of the "basin-formed" geometry. Because the converging force at the end of the Triassic in the eastern (Songpan-Garze) section between the northern and southern plates was mainly in the N-S direction, there formed in the region between the Anymaqen belt southward and the Danba-Yajiang line the mainly E-W arcuate deep- and shallow-level detachment-thrust system, in the front of which there are stacking thrust sheets of pre-Triassic metamorphic series (Neoproterozoic-Triassic) and in the rear there occurred Indosinian syn-tectonic granite (181-199Ma) [1,12]. In the southwest of the Songpan-Garze terrain and the Batang-Litang belt, the Indosinian crustal deformation is characterized by a series of eastward nappes, among which, the "Zhongza" nappe stacking on the Yidun arc came from the Qiangtang terrain to the west [12]. In the western (Bayan Har) section, because the plate converging at the end of the Late Triassic was mainly in the direction of NNE-SSW, the Bayan Har flysch series (Early-Late Triassic) was compressed into tight WNW-ESE steep folds with parallel cleavage which was regionally , arranged in the shape of fan. Along with the folding were low-grade metamorphism (lower green schist facies) and the development of numerous thrusting faults, among which, the Meiyaogou fault to the south of the South Kunlun fault is the greatest ductile reverse fault.

(C) Indosinian Crustal Deformation of the Tanggula-Qiangtang Terrain

The Indosinian deformation in the Qiangtang-Tanggula terrain got weaker southwards from the Jinshaiang suture. The deformation structure of the Jinshajiang suture is characterized by a series of large-scale thrusting shear belt indicated by the high-strain mylonitized rocks. The shearing vergence in the northern section is northward and that in the southern southward, forming a typical radial "Kober" structure cut by a series of Cenozoic strike-slip and thrusting faults. In the inner volcanic belt (the Batang Group of Upper Triassic) in the continental margin of the Qiangtang terrain south of the suture, there developed east-west folds with vertical cleavage.

(2) Tanggula Orogeny

The 500m-thick Middle-Upper Jurassic shallow sea-neritic sediments are the latest marine record on the uplifting. Because the Tanggula orogeny caused by the collision between the Tanggula-Qiangtang terrain and Gondwana at the end of the Jurassic deformed the Middle-Late Jurassic strata on the terrain, forming 300km-wide Alpine "Jura-type" concentric folds and thrusting faults, the crust was shortened about 35% (Fig. 7).

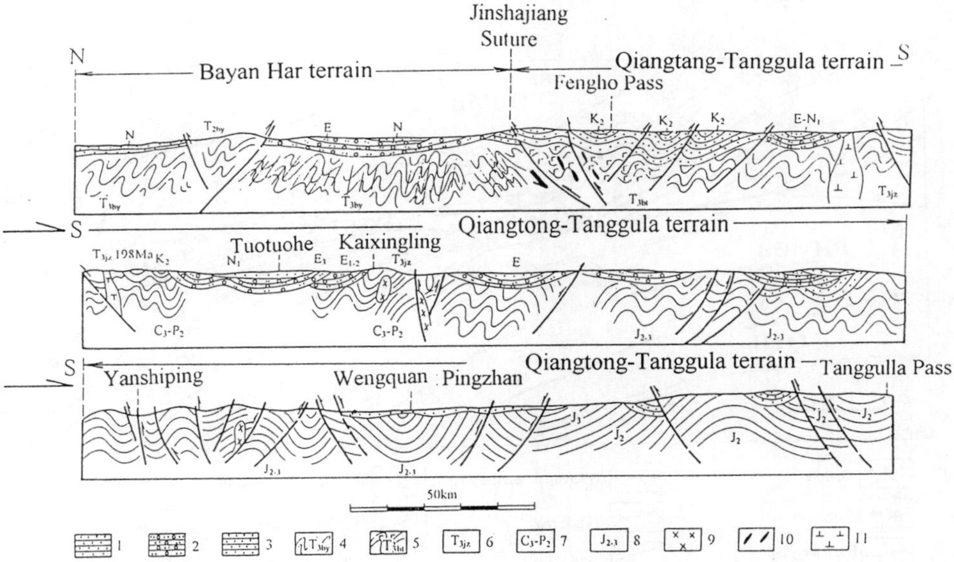

Figure 7. Bayan Har-Tanggula structrual section. 1. Neogene sediments; 2. Paleogene sediments; 3. Upper Cretaceous continental sediments; 4. Upper Triassic Bayan Har Group; 5. Upper Triassic Batang Group; 6. Upper Triassic JizhaGroup; 7. Upper Carboniferous-Upper Permian; 8. Middle-Upper Jurassic; 9. gabbro; 10. ultrabasic rock; 11. diorite

LARGE-SCALE DUCTILE STRIKE-SLIPPING AND BLOCK ROTATING

The Anymaqen structural belt revealing the oceanic crust of the Paleotethys between the East Kunlun and the Bayan Har-Songpan-Garze terrains shows different features in the deformation model and mechanism in its different sections, the western section (west of the Buqingshan Mts.) characterized mainly by sinistral strike-slipping, the middle section (between the Buqingshan Mts. and Huashixia) by oblique thrusting and the eastern section (Huashixia-Maqin) by southward thrusting.

The 5-6km-wide mylonite belt and high-shear strain rocks constitute the bulk of the middle-western section of the large-scale ductile shear zone. In the syn-tectonic granites associated with the shear zone, some are stripped and parallel with shear zone and the others occur to the north with their long axis striking N65 ° W, in the shape of spindle, water drop and ball oblique to the shear zone. In contrast, in association with the strike-slipping in the south are the fold bundles striking about N65 — 60 °W, sinistral en enchlon, in the Triassic Bayan Har flysch series, with no syn-tectonic granite (Fig. 8).

(1) Characteristics of shearing strain and shearing mechanism
The rocks in the shear zone are mainly granitic mylonite, phyllonite, gneiss, micca schist, amphilole schist, etc. The foliation is nearly vertical, the streching lineation is horizontal and the rotation strain and the C-axis fabrics of quartz suggest sinistral strike-slipping. Subsequent brittle deformation superimposed on the ductile deformation and there developed breccia along the zone. Pull-apart basins, en-enchlon earthquake mound and pond and the displacement of river terrace all suggest sinistral strike-slipping.

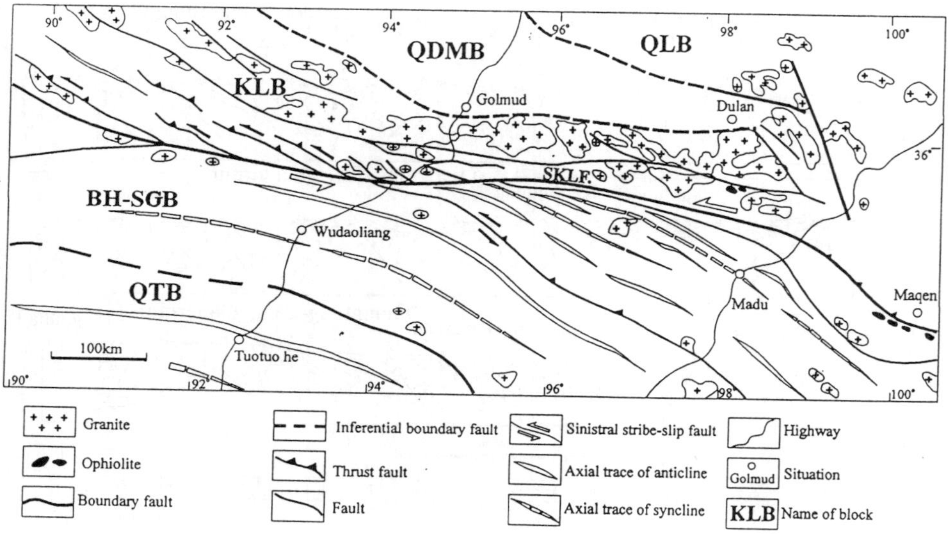

Figure. 8. Strike-slip fault belt of the southern Kunlun and structural map of both sides. QLB — Qilian block; QDMB — Qai dam block; KLB — Kunlunblock; BH — SGB — Bayan Har — Songpan Garze block; QTB — Qiang Tang block; SKLF — Sonthern Kunlun Fault

Calculation of the rate of converging R (R=amount of converging/amount of strike-slipping) [7] reveals that the western section had a converging strike-slipping mechanism (R=0.58) and the middle section had an oblique shortening mechanism (R=1.09) and the main stress was in the direction of NNE-SSW [4,5].

(2) Strike-slipping and syn-tectonic granites
The granite in the strike-slipping zone and to the north are all syn-tectonic granites produced by shearing. The main reasons are:

(A) As long strips, the granites show obvious, strong ductile flowing deformation and shearing and intruded the Early Paleozoic sandy slates and schists. The en-enchlon granites to the north have a geometry of converging toward the shear zone and vertical flowing plane parallel to their long axes and horizontal flowing structures. Deformation is stronger near the center and weaker beyond.

(B) There is obvious migmatization. The granites are transitional to their country rocks which also suggest sinistral strike-slipping.

(C) In the granites there are xenolith of the country rocks that had been sheared.

(D) Petrochemical analysis indicates that the granites have the features of crustal S-type granite, $\delta^{18}O > 8$ ‰, initial $^{87}Sr/^{86}Sr$ ratio >0.708, HREE depleted, LREE enriched, Nb depleted among the trace elements, and Th enriched, resulting from intra-continental deformation.

(3) Age and displacement of the strike-slipping

Displacement of strike-slipping is calculated on the parameters of shearing strain in the shearing zone of different kinds. In the weak strain zone (3km wide) in the marginal portion of the displacement ductile shear zone, the shear angle Q (S∧C) averages 15°, shearing strain (r)=2/tg2Q=3.5, and the displacement is about 10km; In the strong strain zone (0.5-2km wide) in the central portion, r=35 and the displacement is about 70km.

Quantitization of the fold system in association with the strike-slipping gives the direction of shortening in the eastern section (in the vicinity of Huashixia) of the strike-slip fault as 30°-45°, the shortening rate in the NE-SW direction 70-80% and that in E-W direction about 15%; and in the western section (at a point of the Kunlun Pass), the direction of shortening as 20°-30°, shorterning rate 60% [4,5]. The ages (220-240Ma, 150-140Ma, 120-100Ma and 20Ma) of the syn-tectonic granites in association with the strike-slipping, which can basically represent the age of the strike-slipping, indicate that the ductile strike-slipping began in the Middle-Late Triassic and lasted until 20Ma, and since then it has been brittle strike-slipping.

CRUSTAL STRUCTURE, EVOLUTION AND DYNAMICS IN THE MESOZOIC

(1) Shape of the Mesozoic mountain chain and reconstruction of crustal structure

The East Kunlun, Bayan Har-Songpan-Garze and the Tanggula-Qiangtang terrains collided at the end of the Late Triassic and formed an integrated composite mountain chain, but they had their own shapes of mountain chain and features of crustal structure (Fig. 9).

A. East Kunlun Mountain Chain

Because the subduction of the Anymaqen ocean and the subsequent collisional orogeny shortened and thickened the crust of the East Kunlun and resulted in the southward polarity, there formed thrust stacking mountain chain. There occurred in the terrain in succession subduction island arc volcanic rocks, granite, volcanic rocks and granites of continental margin, collision granites and post-orogeny volcanic rocks and granites, and all these show that the crust of the East Kunlun had a high geothermal gradient and was warm.

B. Bayan Har-Songpan-Garze mountain Chain

Since the enormously thick Triassic flysch series of the Bayan Har group and the Xikang group appeared as a "soft cushion' of passive margin on the hard basement (pre-Triassic) of the western margin of the Yangtze block, there formed a "soft cushion-hard basement" crustal structure. Because of the divergent subduction of the

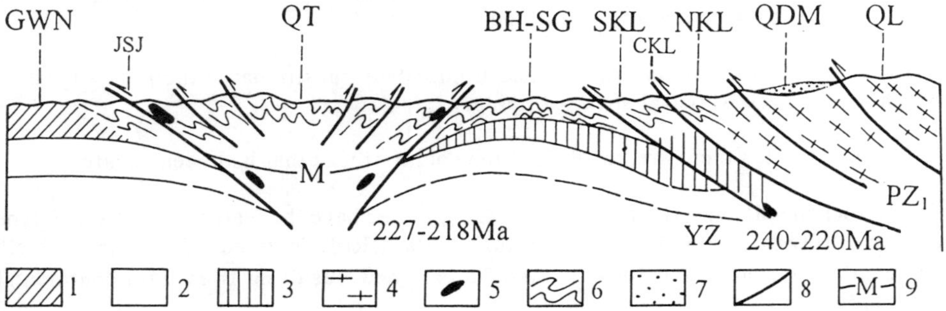

Figure 9. Generalized crustal section of the East Kunlun-Tanggula Mesozoic Composite Mountain Chains. 1. Gondwanaland; 2. manlle; 3. basement of Yangze block; 4. early-paleazoic folded belt; 5. Ulrra mafic rocks; 6. fold; 7. meso-cenozoic basln; 8. thrust; 9. infered Moho

Paleotethys (the Anymaqen and Jinshajiang oceans), the crust of the eastern (Songpan-Garze) section of the terrain had a "hetero-axis, bilateral" orogenic polarity (i.e. southward and eastward), forming "detachment-thrust" mountain chain [12], whereas that of the western section (the Bayan Har) had a "co-axis, bilateral" orogenic polarity (i.e. northward in the north and southward in the south). The strike-slipping in addition changed the mountain chain into "compression-transformation" type. Studies show that very little granite intruded into the upper soft cushion of the crust, in the lower hard basement, however, there occurred high-temperature ductile detachment shear zone and associated local melting in the crust, and thus it is postulated that the crust of the terrain in the Mesozoic was generally cool with warm, viscous low-velocity layers in the depth. The crust of Bayan Har had been consumed under the Kunlun and Qiangtang terrains in the early stage of the Indosinian subduction, gradually forming the crustal cumulate wedge.

C. Tanggula-Qiangtang Mesozoic mountain chain

The "co-axis, bilateral" subduction of the Jinshajiang Paleotethys ocean and the Bangong-Nujiang Neotethys ocean under the Tanggula-Qiangtang terrain produced the following features of the crust:

① The matching and superimposing of the Indosinian orogenic belt (the Late Triassic) in the north and the Tanggula orogenic belt (the Late Jurassic) in the south.

② The crust having a bilateral "Kober" structure (fan-shaped), shortened in the horizontal N-S direction and elongated and thickened vertically.

(2) Formation and evolution of composite mountain chains

Summarizing the above, we have the following stages of the Mesozoic crustal evolution of the East Kunlun-Tanggula Mesozoic composite mountain chain:

A. The Paleotethys (the Anymaqen and Jinshajiang oceans) opened in the the latest-Paleozoic-Early Triassic, separating the Kunlun, Bayan Har-Songpan-Garze and Tanggula-Qiangtang terrains. It is postulated that the Anymaqen ocean was a small shear oceanic basin.

B. The Paleotethys subducted during the Early to Late Triassic. The Anymaqen ocean subducted northward under the Kunlun terrain, forming the Anymaqen subduction system composed of the subduction rock complex, accretionary prism, island arc, back-arc basin and continental margin volcanic arc. On the passive margin of the Yangtze block there formed the "soft cushion" of accretionary rock complex consisting of the Early-Late Triassic flysch series. The Jinshajiang ocean subducted southwards at a larger angle under the Qiangtang-Tanggula terrain, resulting in an extension setting in the northern margin of the terrain and producing the Jinshajiang subduction system characterized by an Andes-type volcanic arc of active margin.

C. The three terrains collided at the end of the Triassic, forming the 500km-wide northern Indosinian orogenic belt.

D. The Bangong-Nujiang ocean opened in the Jurassic, separating the Tanggula-Qiangtang terrain and Gondwana.

E. The Tanggula-Qiangtang terrain collided with Gondwana at the end of the Jurassic, forming the southern Tanggula orogenic belt. The East Kunlun-Tanggula composite mountain chains formed.

F. Since the Cretaceous, the East Kunlun-Tanggula region is in uplifting.

(3) The formation mechanism of the inverted triangular-shaped geometry of the Bayan Har-Songpan-Garze mountain chain

The inverted triangular-shaped geometry of the Bayan Har-Songpan-Garze mountain chain in the Kunlun-Tanggula composite mountain chain has provoked special attention. This mountain chain, which is called the geological Bermuda in China, had gone through the stages of the Indosinian orogeny, Yanshanian post-orogeny and the Cenozoic re-orogeny and had formed, in the Indosinian period, the embryonic form that was a long strip in the west and an inverted triangle in the east.

Studies have shown that the N-S converging between the northern and southern plates in the eastern section of the composite mountain chain formed the Indosinian South Qinling folding-decollement mountain chain and the Indosinian Songpan-Garze folding-decollement mountain chain [11 — 13], whereas in the western section, the (NE-SW) converging between the Qiangtang and East Kunlun terrains resulted in the large-scale South Kunlun strike-slip fault and made the material of Bayan Har terrain flow eastward and caused in the western margin of the Yangtze block nearly E-W compressing, that, together with the N-S converging, produced the nearly E-W arcuate Danba-Yajiang detachment-thrust system.

The NW-SE Xianshuihe strike-slip fault formed in 20Ma cut the Danba-Yajiang system with a sinistral displacement of 80km [12], resulting in the peculiar inverted triangular geometry in the eastern section of the Bayan Har-Songpan-Garze mountain chain.

REFERENCES

1. S.Calassou. Etude tectonique d'une chaine de decollement (Ouest Sichuan et East Tibet, China). *These de doctorat*, Montpellier Ⅱ. 1-169(1995).
2. J. W. Cui, H.zhu, C.D.Wu et al. *Deformation and dynamics of the lithosphere in Qinghai-Tibet plateau*. Geological Publishing House, Beijing.(1992).
3. C. F. Jiang, J.S.Yang, B. G. Fong et al. *Opening-closing tectonics of Kunlun Mountains*, Geological Publishing House, Beijing.(1992).
4. H.B.Li, Z .Q. Xu and W.Chen. Deformational features and tectonic evolution of the South Kunlun strike-slip shear zone, East Kunlun Mountains, *Acta Geoscientia Sinica,* Special issue,. 16-

21.(1996).
5. H. B .Li. The Strike- slip tault at the Southern marg in of East Kun lun orogen and the accompanying granitic magma activity. *Master dissertation*, China University of Geology. 1-94(1996).
6. P.Molnar, et al. Geologic evolution of Northern Tibet: results of an expe dition to Ulugh Muztagh. *Science*, **235**, 299-305 (1987).
7. R. W. Krantz. The transpressional strain strain model applied to strike-slipe, oblique-convergent and oblique-divergent deformation. *J. Struct. Geol.* **17**, 1125-1137 (1995).
8. A.M.C.Sengor. The Cimmeride orogenic system and the tectonics of Eurasia, *The Geological Society of America, Special Paper*, **195**, 82 (1984).
9. A.M.C.Sengor. The Tethyside orogenic system: an introduction; in Sengor, A.M.C. (ed.), *Tectonic Evolution of Tethyan Region*, . 1-22.(1989).
10. X.C.Xiao. *Tectonic evolution and uplift of the Himalaya-Tibet plateau:* 10th Himalaya-Karakoram-Tibet Workshop: ETH Conference Monte Verita (Ascona), Canton Ticino Switzerland. Abstract, pp. 45-46 (1995).
11. Z.Q.Xu. Large shear zones in the main orogenic belts, China, *Episodes*, **18**, 41-43.(1995).
12. Z.Q.Xu, L.W.Hou, Z.X.Wang. *Orogenic processes of the Songpan-Garze orogenic belt of China*: Geological Publishing House, Beijing, 1-190.(1992).
13. Z.Q.Xu and J.W.Cui. *Deformation and dynamics of continental mountain chains:* Metallurgic Publishing House, Beijing, **246**.(1996).
14. Z.Q.Xu, J.S.Yomg and F.Y.Chen, The Anymaqen suture belt and the dynmics in subduction and collision. In: *Study on ophiolites and Geodynamics,* edited by Zhang Qi, Geological Publishing House, Beijing, China, 185-189.(1996).
15. Z.Q.Xu, M.Jiang and J.S.Yang. Deep Tectonophysical Process of the Uplift of the Northern Qinghai-Tibet Plateau--------Evidemce from the lntegrated Geological-Geophysic al Profile from Golmud to the Tanggula Mountains, Qinghai Province, China. *ACTA GEOLOGICA SINICA* **17**, 1-14 (1997).
16. J.S.Yang, P.T.Robinson, C.F.Jiang and Z.Q.Xu. Ophiolite of the Kunlun Mountains, China and their tectonic implications. *Tectonophysics*, **258**, 215-231.(1996).
17. J.S.Yang, X.H.Zheng, W.J.Bai and Y.F.Ren. A preliminary study on genesis of the Durngoi massive Cu-Co-Zn sulfide deposit hosted by the peridotite of A'nyemaqen ophiolite, Kunlun Mt., China, *Proceedings for the 30th IGC.* (in press).(1997).
18. Y.F.Zhang amd J.K.Zheng. *Geological overview in Hoh Xil, Qinghai and adjacent areas:* Seismic Publishing House, Beijing, **177.**(1994).
19. G.Zhang, Z. Yu, Y. Sun et al. The major suture zone of the Qinling orogenic belt. *Journal of Southeast Asian Earth Sciences.* **3:** 63-76 (1989).
20. N.W. Wang. On the Palaeobiogeography and Plate tectonics of Qinghai-Tibet Plateau. *Bullein of the institvte of Geology,* CAGS. Geological Publishing House, Beijing. **9**: 1-28 (1984).

Proc. 30th Int'l Geol. Congr., *Vol.7*, pp. 21-38
Xu Zhiqin, Ren Yufeng and Qiu Xiaoping (Eds.)

Thermo-Mechanical Modeling of Late-Orogenic Extension - A Study from the Variscan Orogen in Central Europe

ANDREAS HENK
Institut fur Geologie, Universitat Wurzburg, Pleicherwall 1, 97070 Wurzburg, Germany

Abstract

A two-dimensional thermo-mechanical finite element model is used to study the post-thickening evolution of the Variscan orogen in Central Europe along a lithosphere-scale transect from the internal zone to the undeformed foreland. The study concentrates on a quantitative evaluation of the processes controlling late-orogenic extension and destruction of crustal roots, particularly the gravitational instability of thickened crust. Modeling results suggest that gravitational forces alone cannot reproduce the observed timing and amount of Permo-Carboniferous crustal thinning. Tensile plate boundary forces are required in addition to gravitation to restore a crustal thickness of about 30 km in the Variscan Internides. Stresses as little as 10 MPa result already in up to 35 % extension and a good fit between model predictions and observed data. It is concluded that the Stephanian to Rotliegend evolution in the vicinity of the modeled traverse resulted not from gravitational forces inherited from Variscan crustal thickening but was related to a modification of the plate boundary stresses at the end of the Westphalian.

Keywords: numerical simulation, gravitational force, extension, crustal thickening, stress

INTRODUCTION

The syn- to late-Variscan evolution of Central Europe between the Visean and Rotliegend (350-255 Ma) was characterized, among others, by rapid exhumation of metamorphic complexes and the development of numerous Late Carboniferous to Permian extensional basins. No Variscan crustal roots are preserved and the present-day crust, undisturbed by Tertiary rifting and Alpine orogeny, shows a uniform thickness of about 30 km [4]. Several authors [22, 23, 12, 24, 13, 6] have drawn parallels to the Tibetan Plateau, Altiplano and Basin and Range Province, suggesting that gravitational collapse of thickened crust was the dominant process controlling the late-Variscan evolution. It has long been recognized that regions of isostatically compensated, thickened crust can be in a state of extensional deviatoric stress [2, 30, 12], but several other processes can contribute to syn- and post-thickening extension in mountain belts. The ones considered most important are syn-convergent lateral extrusion, gravitational instability of the thickened crust induced by topography and thermal relaxation, asthenospheric upwelling and removal of the lower part of the thickened lithosphere by delamination or convective erosion. In addition, changes in plate boundary stresses can significantly enhance late-orogenic extension.

This study was stimulated by work of Braun and Beaumont [5] and Govers and Wortel [16] who used two-dimensional finite element models to simulate lithospheric extension in orogenic settings. Their work concentrate on hypothetical, Himalaya-type orogens with 70 km thick crust resulting from instantaneous, homogeneous thickening. The present study applies similar modeling techniques to the specific geodynamic evolution of the Variscan orogen taking into account thermal relaxation during convergence and crustal thickening. The main objectives are (a) to develop a numerical model which can reproduce the first-order characteristics of late- Variscan evolution in Central Europe and (b) to determine the factors controlling post-thickening extension. The underlying question is if the late-Variscan evolution was exclusively related to the preceding Variscan orogeny or if other processes in addition to simple gravitational collapse affected the geodynamic evolution of Central Europe during the Permo-Carboniferous.

VARISCAN EVOLUTION OF THE STUDY AREA

The evolution of Europe during Devonian to Permian times was largely controlled by the approach and collision of Gondwana, Baltica and several microplates. Continental collision resulted in crustal thickening by thrusting and stacking, ultimately forming the Variscan orogen [14, 13, 15]. Subduction of oceanic crust in all Variscan basins was most likely completed by the end of the Devonian. The geodynamic evolution from the Early Carboniferous onwards can be considered as the collisional stage. Convergence terminated at about 325 Ma in the orogen's interior but lasted until 305 Ma in the most external parts of the Variscan orogen [31]. In the internal zone, crustal collision was contemporaneous with and immediately followed by rapid exhumation of crystalline complexes, voluminous granitic intrusions and high temperature/low pressure metamorphism [13, 21]. The significant amount of syn-convergent exhumation in the Variscan Internides has important implications concerning the crustal thickness at the end of the convergence stage as crustal thickening during continental collision was partly counterbalanced by syn-convergent erosion and particularly by tectonic denudation.

Syn- to post-convergent extension ultimately resulted in the formation of numerous SW-NE trending basins [23]. Although local subsidence is already documented from the Westphalian A onwards, widespread basin evolution did not start before the Stephanian [31]. At the end of the Paleozoic, most of the previously thickened crust in the Variscan realm had regained a thickness of about 30 km. Locally, the crust had even subsided below sea level, as is suggested by the rapid Zechstein transgression over part of the former orogen [23, 12, 31].

NUMERICAL MODEL

Finite element techniques are used to study lithospheric deformation and temperatures during late-Variscan times. The numerical simulation is based on the finite element software ANSYS (registered trademark of Ansys Inc., Houston). It is assumed that the physical properties of the Variscan lithosphere were relatively constant along strike, so that the orogen can be adequately described by a two-dimensional approach. This plane-strain assumption is certainly a simplification regarding the complexity of real orogens and orogenic processes, but it is considered as an important first step towards more sophisticated three-dimensional models.

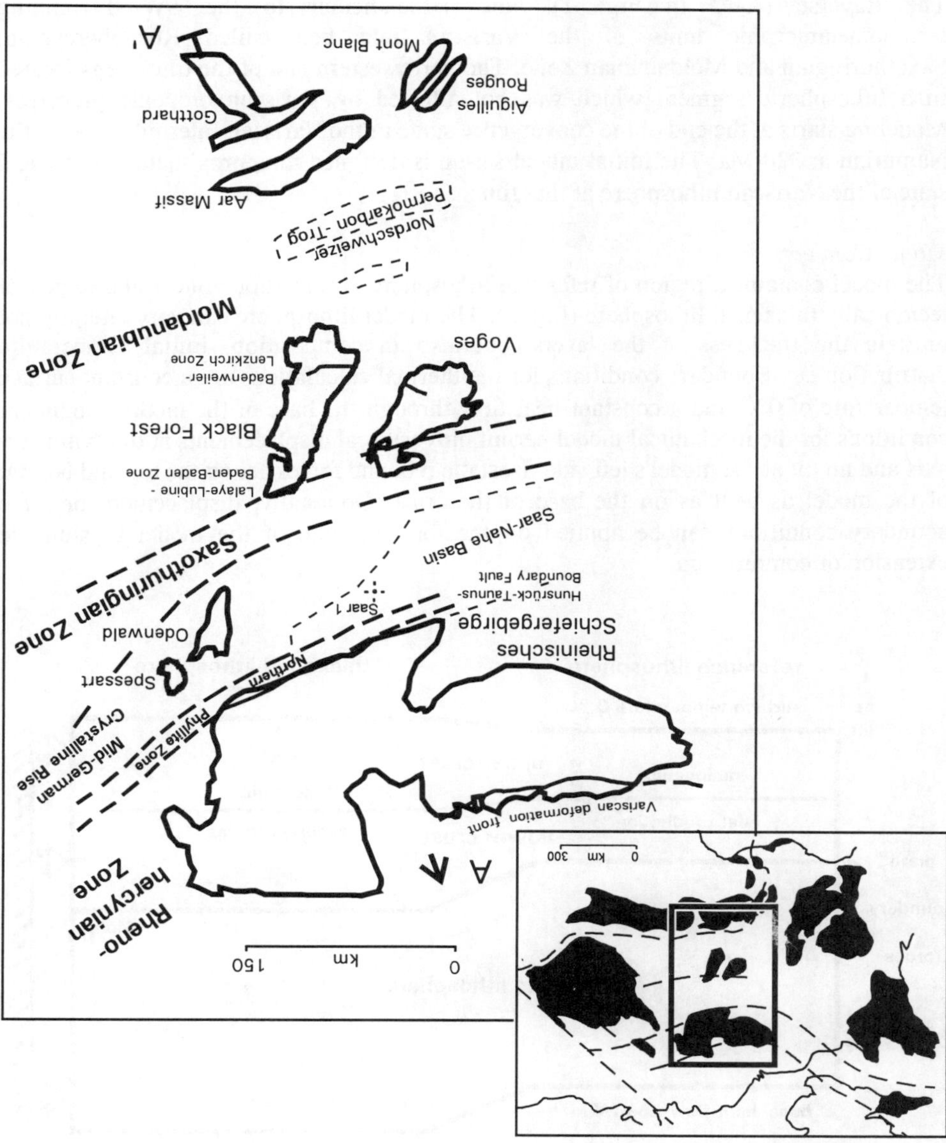

Fig. 1 Simplified geological map of Variscan structures and Permo-Carboniferous basins in Central Europe (modified after Franke, 1988 and Pfiffner, 1993). The Variscan basement units in the Alps are shown in their pre-Mesozoic position. A-A' marks the modeled traverse from the internal zone of the orogen to the undeformed foreland.

The numerical simulation concentrates on a cross-section through the western part of the Central European Variscides which provides a half-profile from the internal zone of the orogen to the undeformed foreland (Fig. 1). Most of the other half is obscured by the Alps and relatively little is presently known to constrain a numerical modeling approach. The traverse (A-A' in Fig. 1) runs perpendicular to the SW-NE-trending tectonometamorphic units of the Variscan fold belt called Rhenohercynian, Saxothuringian and Moldanubian Zone. The northwestern end of the traverse is located in a lithospheric segment which was not affected by Variscan orogenic processes. Modeling starts at the end of the convergence stage in the Variscan Internides, i.e. in the Namurian at 320 Ma. The initial model set-up is designed to approximate the physical state of the Variscan lithosphere at that time.

Model Concept

The model contains a region of reference lithosphere, a transition zone and a region of tectonically thickened lithosphere (Fig. 2). The model lithosphere can vary laterally not only in the thickness of the layers, but also in composition, initial temperature distribution etc. Boundary conditions for the thermal calculations are a constant surface temperature of 0°C and a constant heat flow through the base of the model. Boundary conditions for the mechanical model permit only vertical displacements at the symmetry axis and no tilt at the model's left side. Isostatic rebound forces act on the top and bottom of the model as well as on the base of the crust. Optionally, displacement or force boundary conditions can be applied on the foreland side of the model to simulate extension or compression.

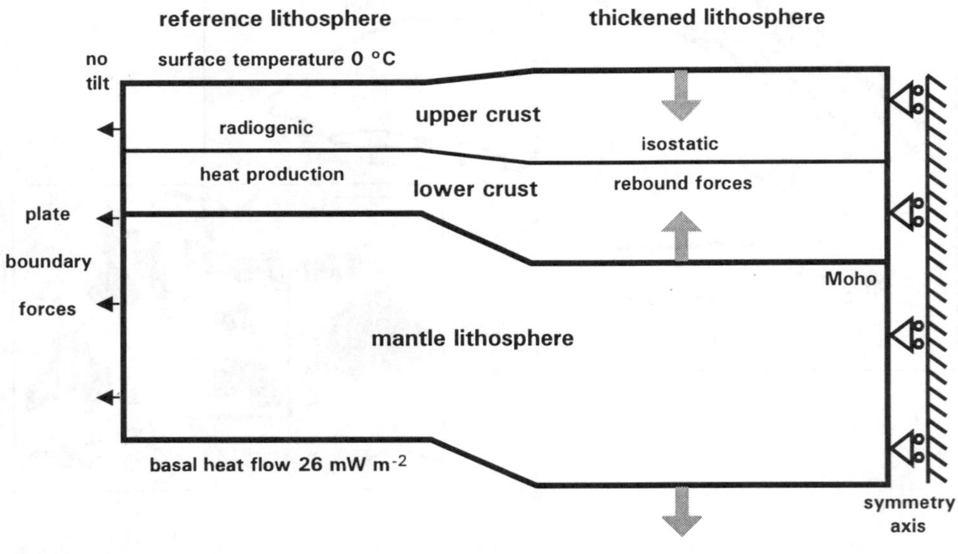

Fig. 2 Conceptual set-up for thermo-mechanical finite element model (see also Braun and Beaumont 1989 and Govers and Wortel 1993).

Thermo-mechanical coupling is modeled with two geometrically identical finite element grids which are used to solve the thermal and mechanical equations successively. Both finite element grids are symmetric about a vertical axis in the orogen's interior (Fig. 2). Thus, only the left half of the orogen needs to be modeled. Calculations are coupled via temperatures and thermal stresses, which result from thermal expansion, and via displacements.

Tab. 1 Thermal and mechaniocal material parameters.

Thermal Lithosphere Properties			
surface temperature (°K)	Ts	273	
asthenosphere temperature (°K)	Ta	1623	
lithospheric basal heat flow (W m^{-2})	q*	0.026	
Upper Crust			
thermal conductivity (W m^{-1} °K^{-1})	k	3-1.3	Zoth 8 Hänel 1988
specific heat ($m^2 s^{-2}$ °K^{-1})	C	1300	
radiogenic heat production (W m^{-3})	H	2.6×10^{-6}	Cermák 1995
Lower Crust			
thermal conductivity (W m^{-1} °K^{-1})	k	2.5 - 1.7	Zoth Hänel 1988
specific heat ($m^2 s^{-2}$ °K^{-1})	C	1300	
radiogenic heat production (W m^{-3})	H	0.52×10^{-6}	Cermák 1995
Upper Mantle			
thermal conductivity (W m^{-1} °K^{-1})	k	3-4	Zoth Hänel 1988
specific heat ($m^2 s^{-2}$ °K^{-1})	C	1300	
radiogenic heat production (W m^{-3})	H	0.02×10^{-6}	Cermák 1995
Mechanical Lithosphere Properties			
reference continent thickness (m)		30×10^3	
reference lithosphere thickness (m)		117×10^3	
Poisson's ratio	ν	0.25	
Upper Crust			
density at 273°K (kg m^{-3})	ρ	2800	
Young's Modulus (Pa)	E	0.5×10^{11}	
strain rate coefficient (Pa^{-n} s^{-1})	a_0	1.633×10^{-26}	Paterson &
activation constant (°K)	Q/R	16238	Luan 1990
power law stress exponent	n	3.1	
Lower Crust			
density at 273°K (kg m^{-3})	ρ	3000	
Young's Modulus (Pa)	E	0.8×10^{11}	
strain rate coefficient (Pa^{-n} s^{-1})	a_0	2.063×10^{-23}	Shelton &
activation constant (°K)	Q/R	28747	Tullis 1981
power law stress exponent	n	3.2	
Upper Mantle			
density at 273°K (kg m^{-3})	ρ	3400	
Young's Modulus (Pa)	E	1.5×10^{11}	
strain rate coefficient (Pa^{-n} s^{-1})	a_0	5.248×10^{-25}	Chopra &
activation constant (°K)	Q/R	59899	Paterson 1981
power law stress exponent	n	4.48	

Thermal and Mechanical Properties

The lithosphere is assumed to be made up of three layers: upper crust, lower crust and mantle lithosphere. The thermal and rheological properties applied to these three layers are summarized in Tab. 1. Calculation of the lithospheric temperature field is based on two-dimensional heat transport by conduction and advection. An important component exhumation which amounts to about 16 km in the Spessart. Further information can be derived from subsidence analysis in the Saar-Nahe Basin, located on top of the Saxothuringian basement. They indicate about 35 % of crustal extension between the Westphalian and Rotliegend so that a pre-Westphalian crustal thickness of about 40 km can be inferred [18]. From the various arguments outlined above, it is concluded that 40 km can be considered a good estimate for the crustal thickness at 320 Ma in both the Saxothuringian and Moldanubian Zones. Thrusting of the Mid-German Crystalline Rise over the Rhenohercynian passive margin resulted in a complex crustal structure within the Saxothuringian domain. Refraction seismic data (DEKORP Research Group C 1994) reveal two crustal slices stacked on top of each other. This feature is also incorporated in the initial model geometry (Fig. 3b). No such crustal structure is known from the Moldanubian Zone and, hence, normal crustal layering is assumed.

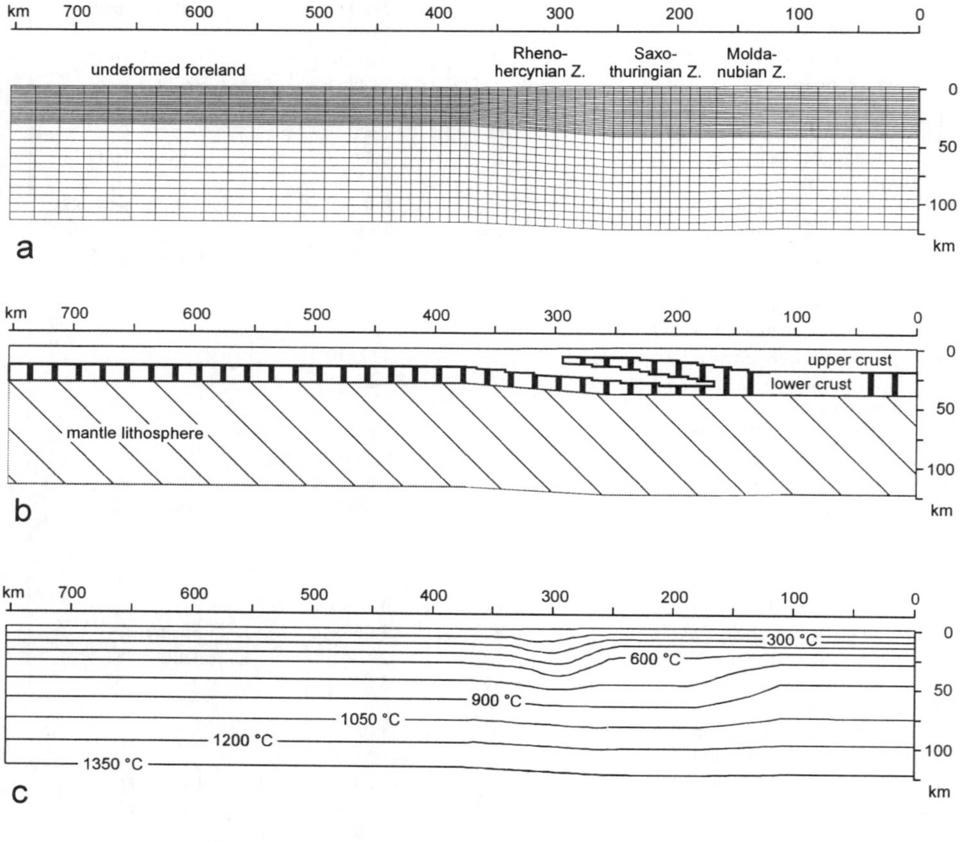

Fig. 3 Initial finite element grid, material distribution and transient temperature field (in °C)

The initial thickness of the orogenic mantle lithosphere is difficult to constrain. Regarding the strength of the syn-convergent thermal peak (e.g. widespread granulite metamorphism and granite intrusions) and the various subduction zones, a rather thin mantle lithosphere in the Variscan Internides seems plausible. It is therefore assumed of the heat flow measured at surface results from the decay of radiogenic elements mainly within the crust. Values for the radiogenic production in the Variscides are taken from Cermák [9]. For the temperature-dependence of thermal conductivity the empirical relationships of Zoth and Hänel [32] are used.

Lithospheric deformation can be characterized by various rheological laws which depend mainly on rock composition, temperature, pressure, strain and strain rate [28]. The most important deformation mechanisms are brittle failure and viscous flow. Consequently, a viscoplastic rheology was adopted for the lithosphere. When the ductile stress is less than the yield stress, the crust and mantle are assumed to behave as incompressible viscous fluids deforming by creep. When the ductile stresses exceed the yield stress, they are assumed to deform according to an perfect plastic flow law. This yield stress is defined by Byerlee's [7] frictional relation which results in a tensile strength gradient of 14 MPa km^{-1} for hydrostatic pore pressures. Stresses in the ductile domain are assumed to be governed by thermally activated power law dislocation creep. The non-linear relationship between strain rate $\varepsilon°$, differential stress σ_1 - σ_3, and temperature T is given by

$$\varepsilon° = a_0(\sigma_1 - \sigma_3)^n \exp(\frac{-Q}{RT})$$

where R is the universal gas constant and a_0 (strain rate coefficient), n (power law stress exponent) and Q (activation energy) are material properties derived from laboratory experiments. The rheological stratification of the Variscan continental lithosphere is modeled using creep parameters of wet quarzite, anorthosite and wet dunite for the upper crust, lower crust and mantle lithosphere, respectively [26, 29, 10]. These creep parameters do not imply a corresponding monomineralogic composition of the Variscan lithosphere but approximate the rheology of the weakest interconnected phase which determines the actual strength of the rock in each of the three layers [17].

Initial lithospheric structure

Reconstructions of the Visean paleogeography [31, 25] show a widespread distribution of marine sediments in the vicinity of the modeled traverse. This suggests that - with the exception of local magmatic arcs - most of the crust in the area of the future orogen had a crustal thickness of less than 30 - 32 km prior to thrusting (see also Eisbacher et al., [13]. Information on the crustal thickness at 320 Ma, i.e. the start time of the numerical simulation, is not readily available but several lines of evidence can be used to constrain the initial model geometry. Stephanian sediments in the Variscan Internides frequently overlie an amphibolite facies metamorphic basement indicating already a substantial amount of crustal thinning by erosion and/or tectonic prior to the Stephanian. For the crustal thickness in the Black Forest area, Eisbacher et al. [13] give an estimate of possibly up to 40 km. This corresponds with evidence from the metamorphic history. Even if the crust was shortened by a factor of two during continental collision, the syn-convergent exhumation of at least 25 km may have counterbalanced major part of the crustal thickening so that probably no more than about 40 km of crustal thickness was left at the beginning of the post-convergent evolution. The maximum crustal thickness in the Saxothuringian Zone was estimated at 55 km from balanced cross sections by Behrmann et al. [3]. This is an extreme value as it does not consider syn-convergent that during collision the mantle lithosphere of the upper plates detached at the Moho and was absorbed without major thickening in the upper mantle of the underthrust crustal segments.

The rapid convergence and exhumation stage immediately before the onset of the numerical simulation strongly disturbed crustal temperatures. Thus, a transient temperature field is required as initial temperature distribution (Fig. 3c). Lithospheric temperatures in the Moldanubian Zone at 320 Ma are calculated by assuming doubling of 32 km thick crust within 10 Ma (350-340 Ma) followed by 10 Ma of exhumation at rates of 2.4 mm/a (340-330 Ma) and 10 Ma of thermal relaxation (330-320 Ma). Similarly, temperatures in the Saxothuringian zone are based on stacking of a 26 km thick crust on top of the reference lithosphere between 330 and 325 Ma followed by 5 Ma of exhumation at rates of 3.2 mm/a. These scenarios can reproduce the characteristics of the available P-T-t data [19]. Convergence and thrusting of the Saxothuringian onto the Rhenohercynian Zone, which occurred at horizontal velocities of 2 cm/a during the Late Viséan and Namurian [19], also modified the temperature field in the foreland. The corresponding temperature distribution in the overthrust structure is taken from Henk [19].

Model Geometry
The numerical model comprises a 750 km long section ranging in thickness from the Earth's surface to the base of the lithosphere (Fig. 3a). The model geometry contains a 240 km long section of lithosphere with 40 km thick crust. The transition zone between normal and thickened crust has a length of 120 km. It corresponds to the Rhenohercynian Zone, whereas the thickened section represents Saxothuringian and Moldanubian Zone as well as the Variscan basement units in the Alps. The thickness of the model lithosphere ranges between 117 and 125 km. Altogether, the model geometry consists of 2108 four-node elements.

Crustal and upper mantle thicknesses of 30 km and 87 km were selected for the thermally stable reference lithosphere. Using the material parameters listed in Tab. 1, the top of the reference lithosphere is set to be close to sea level as indicated by the marine to deltaic sedimentation during the Carboniferous. Additionally, this reference lithosphere is in close isostatic and potential energy equilibrium with midoceanic ridges [20, 30].

MODELING RESULTS

Model A1: Extension of 40 km thick crust by gravitational forces only
This model is designed to study a scenario in which gravitational collapse in the Variscan Internides started right at the end of the crustal stacking phase. Compressive plate boundary stresses are set to zero at the beginning of the thermo-mechanical calculations, i.e. at $t = 0$. Subsequently, only gravitational forces act on the lithosphere. Numerical simulation results indicate an early stage of rapid extension in the Saxothuringian and especially Moldanubian Zone, but extension decreases rapidly after 5 to 10 Ma and is negligible after 75 Ma. The cumulative vertical strain amounts to 6 and 13 % in the Saxothuringian and Moldanubian Zone, respectively (Fig. 4). Thus, at

the end of the simulation crustal thicknesses between 35.5 and 38 km and a surface topography of 690 to 960 m remain in the thickened part of the model (Fig. 5). The lateral differences in crustal thinning are mainly a result of the assumed initial crustal structure. The crustal stack in the Saxothuringian Zone raises lower crustal rocks to shallower crustal levels and hence lower temperatures in comparison to adjacent areas. This increases the overall strength of the lithosphere in the Saxothuringian domain making it less prone to crustal extension. A close examination of the style of deformation in this area reveals differential movement between the two crustal slices which resemble extensional reactivation of a thrust zone. In the numerical experiment, the upper crust of the underthrust plate acts as a low-viscosity zone promoting movement of the hanging wall towards the orogen's interior. Maximum total strain in this zone after 35 Ma is 68 % (Fig. 5c).

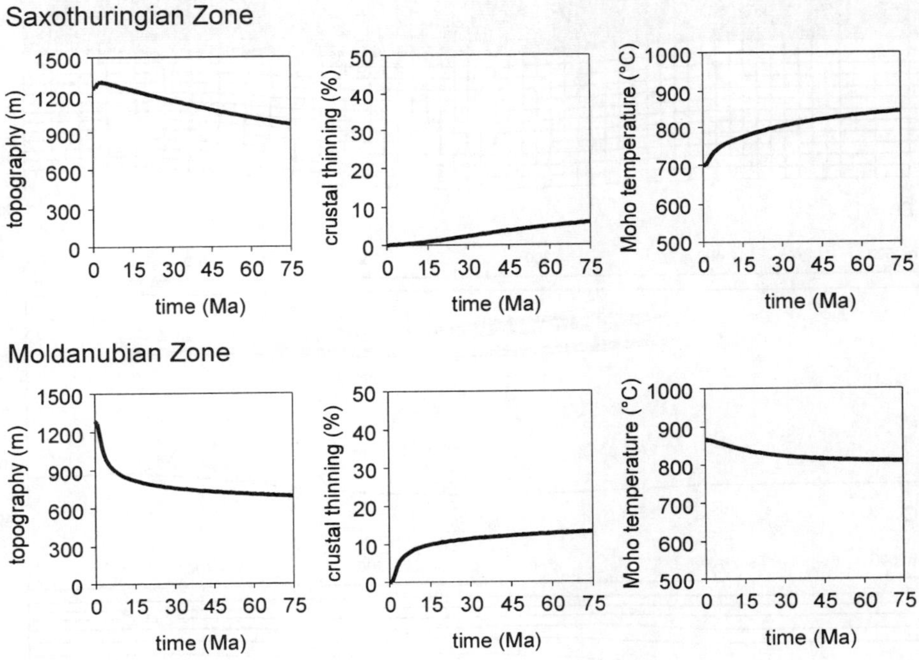

Fig. 4 Model Al: Topography, cumulative vertical strain and Moho temperature as a function of time at the centers of the Saxothuringian and the Moldanubian Zone.

Thermal relaxation in the Saxothuringian Zone increases Moho temperatures from 700 °C to 840 °C slowly approaching equilibrium conditions (Fig. 4). The effect on the lithospheric rheology is a continuous reduction in strength which is documented by a rather constant increase in total strain. The opposite effect can be observed in the Moldanubian Zone, where thermal relaxation prior to 320 Ma was already more advanced. The larger amount of strain there resulted in stronger thinning of the radiogenic layers and temperatures drop from initially 870 °C to 810 °C. This in turn slowly increases the strength of the lithosphere and extension decreases much more

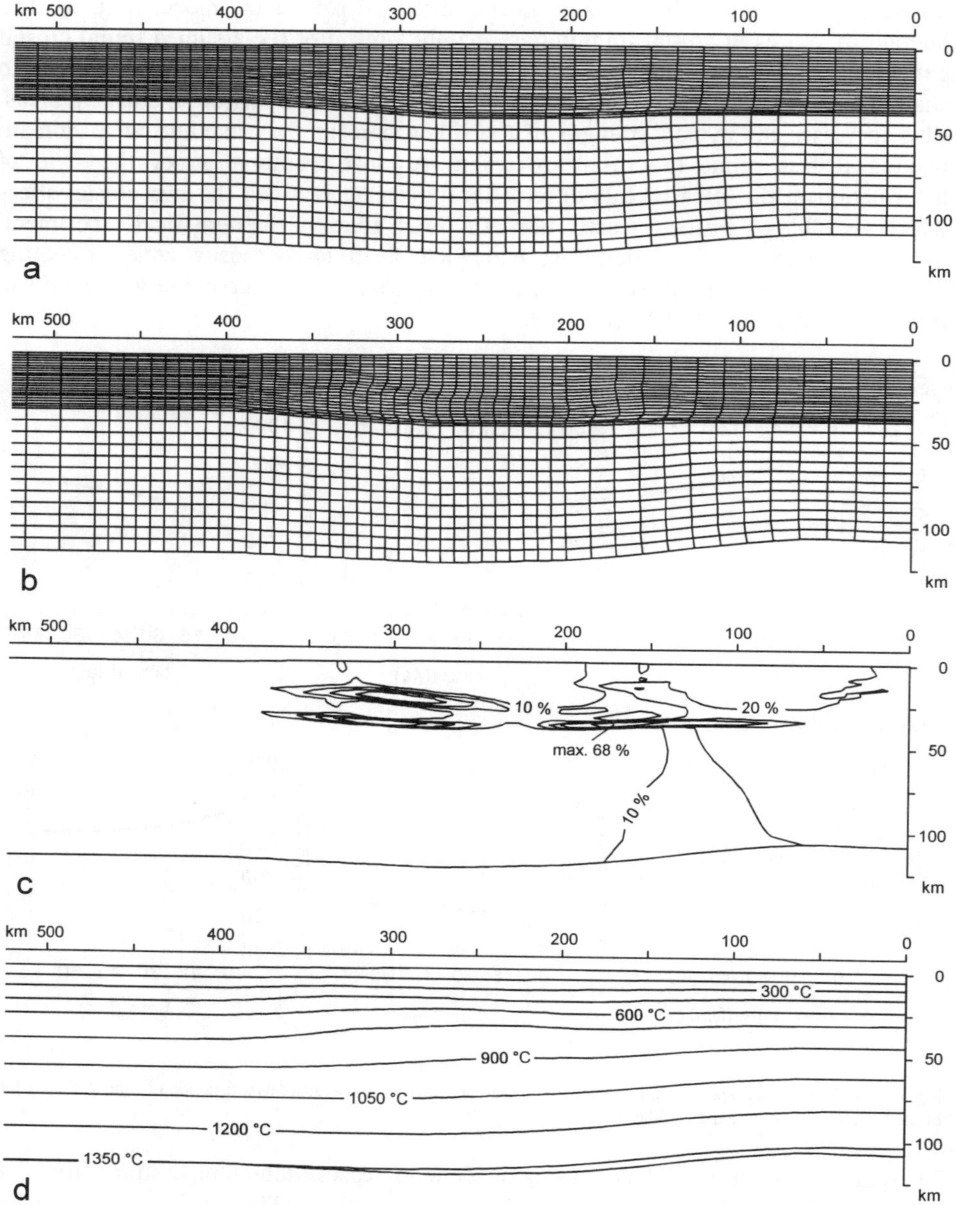

Fig. 5 Model A1: a. FE-grid after 15 Ma. b. FE-grid after 35 Ma. c. Cumulative strain after 35 Ma. d. Temperature distribution after 35 Ma.

rapidly than in the Saxothuringian Zone.

Model B1: Extension of 50 km thick crust by gravitational forces only

Model Bl differs from the previous model set-up in only one respect - an initial crustal thickness of 50 km. In the light of the arguments outlined above this can be considered a maximum value for the Variscan Internides. As before, only gravitational forces act on the lithosphere. Modeling results show significantly more extension but gravitational forces fail again to restore the reference crustal thickness in the internal part of the orogen (Fig. 7). After the initial phase of rapid extension, crustal thinning slows down rapidly and the cumulative vertical strain in the internal zone amounts to 25 - 30 % after 35 Ma (Fig. 6). The surface topography drops from about 2500 m to 950 m, but the crustal thickness is still about 39 km. Due to the large radiogenic heat production, temperatures at the base of the thickened crust rise to 860 °C in the Saxothuringian zone and 930°C in the Moldanubian zone, respectively.

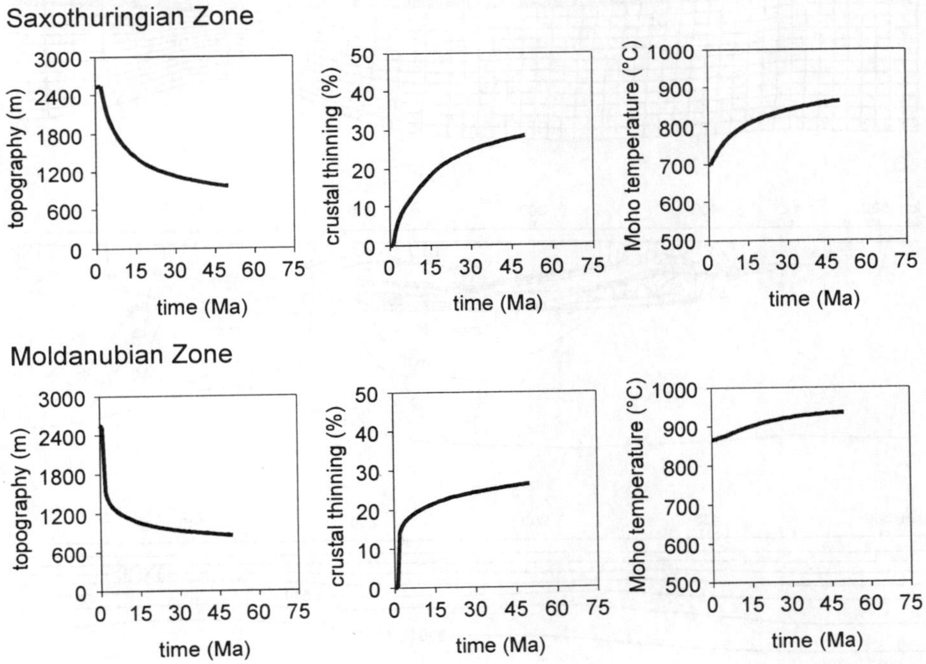

Fig. 6 Model Bl: Topography, cumulative vertical strain and Moho temperature as a function of time at the centers of the Saxothuringian and the Moldanubian Zone.

The strain distribution after 35 Ma (Fig. 7c) shows a number of interesting details. Several dome-like structures have been formed in the lower crust which strongly resemble metamorphic core complexes. They are overlain by high-strain zones which have accumulated as much as 300 % of deformation. There are also lateral differences in the amount of thinning in the crust and upper mantle. The zones of maximum crustal and mantle lithospheric extension do not coincide spatially. Thinning of the mantle

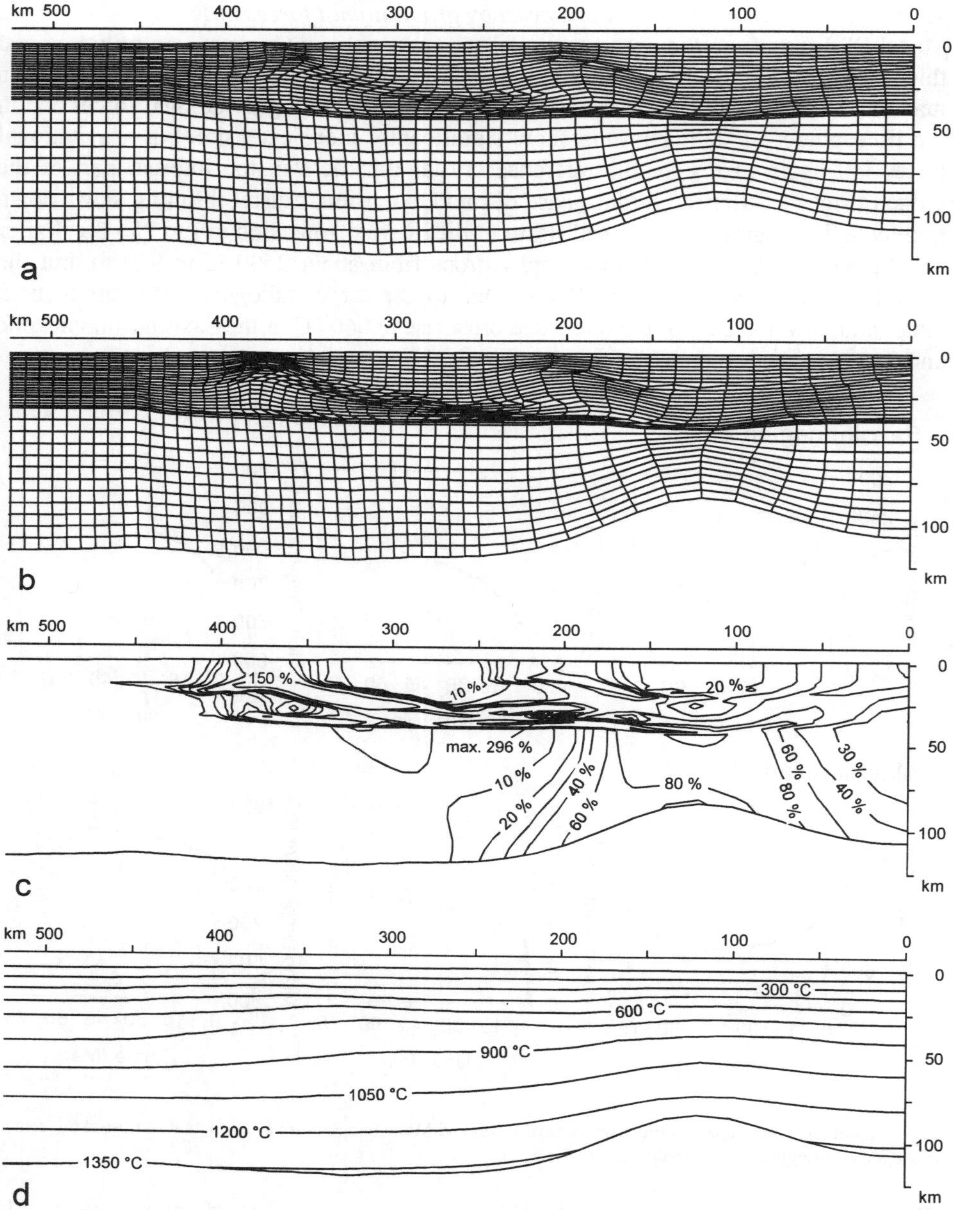

Fig. 7 Model B1: a. FE-grid after 15 Ma. b. FE-grid after 35 Ma. c. Cumulative strain after 35 Ma. d. Temperature distribution after 35 Ma.

lithosphere is greatest where the orogenic crust still has the largest thickness and the Moho its highest temperature, respectively. The reason for this strain pattern is the temperature- and strain-rate dependence of the creep laws in the ductile domain. Already small heterogenities in the temperature and material distribution of the initial model set-up tend to localize deformation. Due to strain-softening effects, these deformation zones can keep their position throughout the numerical simulation.

Model G3: Extension of 40 km thick crust by gravitational forces and tensile plate boundary stresses following 15 Ma of thermal relaxation

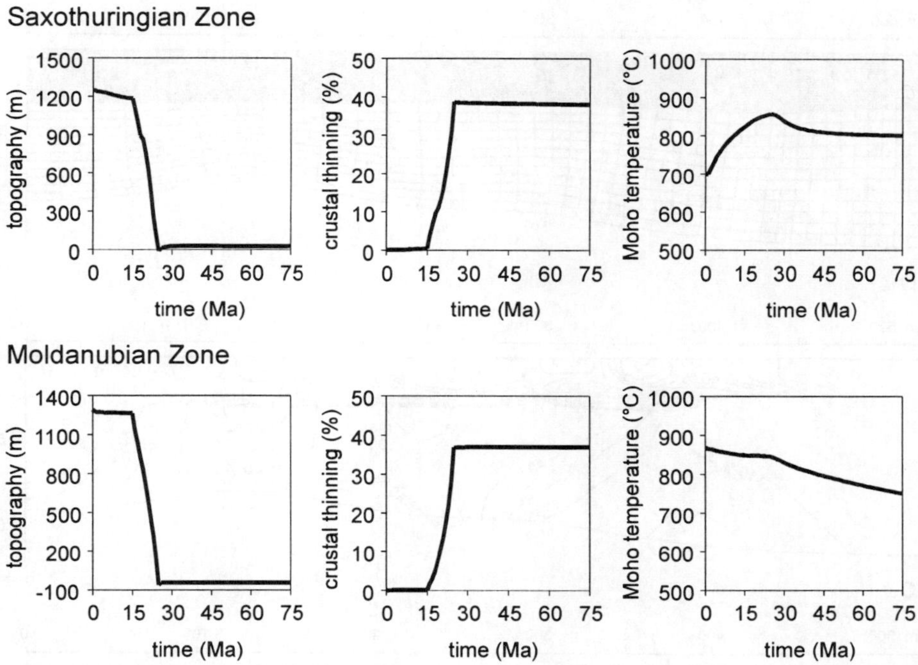

Fig. 8 Model G3; Topography, cumulative vertical strain and Moho temperature as a function of time at the centers of the Saxothuringian and the Moldanubian Zone.

In order to minimize the lateral differences in crustal thinning between the Saxothuringian and Moldanubian Zones, the initial geometry of model G3 does not include a thrust structure like models Al and Bl. Instead, normal crustal layering with a lower crustal thickness of 10 km is assumed. The scenario for model G3 starts with a phase of thermal relaxation, but at t = 15 Ma compression is replaced by tensile plate boundary forces acting on the foreland side of the model. This is achieved by linearily adding a tensile force of 10 MPa between 15 and 25 Ma. Although rather small, the addition of tensile plate boundary stresses dramatically changes the stress balance and deformation pattern in comparison to model Al (Fig. 8 and 9). During the resulting

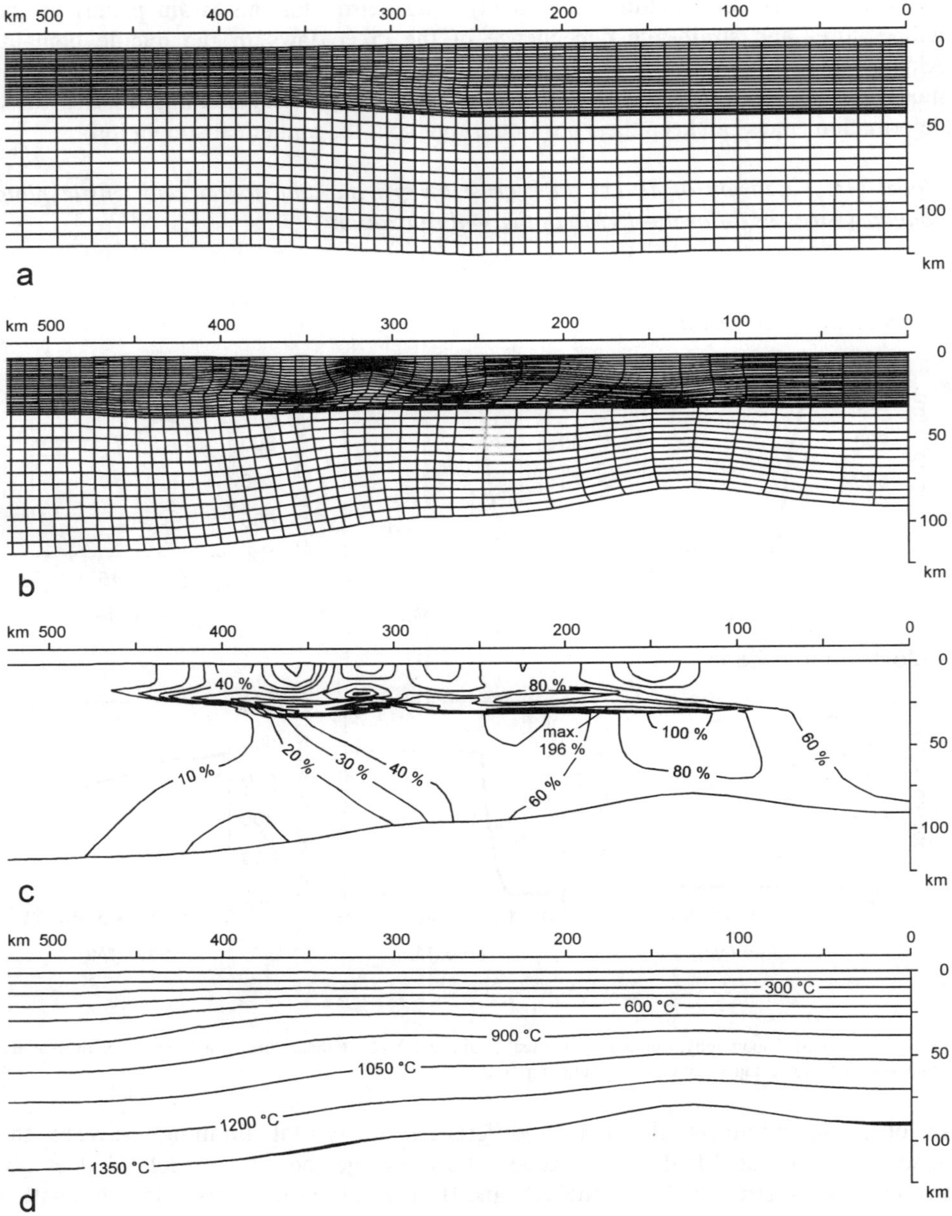

Fig. 9 Model G3: a. FE-grid after 15 Ma. b. FE-grid after 35 Ma. c. Cumulative strain after 35 Ma. d. Temperature distribution after 35 Ma.

rapid extension phase, the cumulative vertical strain amounts to about 35 %. The crustal thickness decreases from 40 to less than 30 km and the surface topography drops to about sealevel. Thus, the crustal roots, which resulted from crustal thickening, are completely destroyed by the tensile plate boundary forces. Due to the large amount of crustal thinning, Moho temperatures in the internal zone decrease to 750°C after 75 Ma.

The strain distribution in the model lithosphere (Fig. 9c) shows again the formation of dome structures in the lower crust. They are overlain by high-strain zones mainly dipping towards the orogen's interior. The most prominent strain zone is the Moho where deformation reaches almost 200 %. This is a result of the lateral variable degree of extension in crust and mantle lithosphere. Locally, thinning of the upper mantle can amount to 70 %, which is twice the extension in the crust. The base of the lithosphere in this area is located at a depth of about 80 km.

DISCUSSION

The numerical experiments are designed for comparison with the late-Variscan evolution of Central Europe. The start time of the thermo-mechanical models is equivalent to 320 Ma and the initial model set-up approximates the state of the Variscan lithosphere during the Namurian. Interesting moments in time to compare the calculation results with the real evolution are t = 15 Ma, the onset of the widespread Stephanian to Rotliegend basin evolution at 305 Ma, and t = 30 Ma, the peak of volcanic activity in Europe at 290 Ma. Active crustal extension and syn-tectonic basin subsidence, respectively, ceased at about 285 Ma which is equivalent to t = 35 Ma.

The results of the numerical experiments strongly suggest that gravitational forces are insufficient to restore a crustal thickness of about 30 km along the modeled lithospheric transect. Models Al and Bl, which rely solely on the gravitational instability of the orogen, fail to reproduce the timing and amount of observed Permo-Carboniferous extension. In spite of the high temperatures, the remaining strength of the lithosphere is still sufficient to balance most of the gravitational potential energy stored in the thickened crust. This is in agreement with the findings of Braun and Beaumont [5] who state that the crust will only return to its reference thickness if the lithospheric strength is close to zero or if forces in addition to gravitation act. Therefore, it seems unlikely that gravitational collapse was the dominant geodynamic process controlling Stephanian to Rotliegend evolution in Central Europe.

Several parameter studies were carried out to assess if this conclusion depends critically on the geometry and material parameters selected for the model lithosphere. The influence of a greater initial crustal thickness was already studied by model Bl. Another important parameter could be the selected flow laws as lower lithospheric viscosities would possibly enhance the gravitational instability. Some of the total lithospheric strength comes from the brittle part of the upper crust for which strength is independent of rock type [7]. Because of the high temperatures, lower crust and upper mantle are already regions of very low strength. The use of creep laws with even lower viscosities only reduces this subordinate contribution to the overall lithospheric strength and no significant effect for the modeling results is found. The same conclusion can be drawn for higher lithospheric temperatures. Such model runs show a decrease of the already very low strength of the lower crust and upper mantle but the strength of the brittle crust is not affected as it is independent of temperature. Higher temperatures raise also the depth of the brittle-ductile transition, but this effect is minimal in the present study because this transition is already at a depth of about 10 km and hardly raises much further. These sensitivity studies suggest, that the general conclusion concerning the relative inefficiency of gravitational collapse in the Central European Variscides holds for a certain range of material parameters and does not depend critically on the specific values selected for the model lithosphere. The numerical simulation results are considered as a conservative estimate in the sense that all conditions, i.e. weak rheologies, high temperatures etc., were already selected in order to support a gravitational collapse.

The results of model G3 illustrate that a crustal thickness in the Variscan Internides of about 30 km can only be re-established within a reasonable time span if tensile plate boundary forces act in addition to gravitation. For a tensile force as little as 10 MPa, the calculated amount of crustal thinning is already in good agreement with the observed present-daythickness. The source of the inferred tensile plate boundary forces in the Variscides could be the reorganisation of the regional stress field at about 305 Ma [31, 6]. At this time, movement between Gondwana and Laurussia changed to dextral translation [1, 31]. A simple decrease of the horizontal compressive forces at the plate boundaries as envisaged by Burg et al. [6] seems to be insufficient regarding the results of thermo-mechanical modeling. The paleogeographic maps of Ziegler [31] illustrate that the Westphalian basin evolution followed the trend of the Variscan structures and was probably still related to late-stage orogenic processes, i.e. tectonic escape [6]. In contrast, the Stephanian to Rotliegend basins evolved partly oblique to the Variscan fold belt which at this time was disected by a complex pattern of mainly NW-SE trending dextral strike-slip zones [1, 31]. These differences also support the view that the orientation of the stress field changed at the transition from Westpalian to Stephanian. Subsequently, a modified far-field stress regime was superposed on the gravitational forces in the Variscan Internides.

Strictly speaking, the conclusions derived from this study are only valid for the vicinity of the modeled traverse. Application to other parts of the Variscides requires careful examination of the specific physical state of the lithosphere prior to Permo-Carboniferous extension. Care should be taken to distinguish between syn-convergent extension, which was confined to upper crustal levels and post-convergent extension affecting the entire lithosphere.

CONCLUSIONS

The main conclusion of the present study is that crustal extension in Central Europe from about 305 Ma onwards resulted most likely not from gravitational forces inherited from Variscan crustal thickening but from tensile forces acting at the plate boundaries. The Stephanian to Rotliegend evolution was therefore post-Variscan in the sense that the main driving force for crustal extension was not genetically related to the preceding orogeny. Modification of plate boundary stresses had such a pronounced effect in the Variscan realm mainly because of its reduced lithospheric strength in comparison to adjacent regions not affected by orogenic processes.

More numerical work is required to extend the modeling approach to three dimensions and to include, for example, the mechanical aspects of the crustal stacking phase as well as the thermal and rheological consequences of Permo-Carboniferous magmatism. Likewise, more field data on the timing and amount of syn- to post- thickening extension in the Variscan fold belt is urgently required to constrain further numerical simulations. But already with the present stage of knowledge, thermo- mechanical modeling proves to be a very valuable tool to study the Permo-Carboniferous evolution of Central Europe on a lithospheric scale.

Acknowledgements

This study benefited greatly from discussions with C. Beaumont, V. Lorenz and O. Oncken. Financial support by the Deutsche Forschungsgemeinschaft as part of the programme 'Orogene Prozesse - ihre Simulation und Quantifizierung am Beispiel der Varisciden' is gratefully acknowledged.

REFERENCES

1. Arthaud F, Matte P. Late Paleozoic strike-slip faulting in southern Europe and northern Africa: result of a right lateral-shear zone between the Appalachians and the Urals, *Bull geol Soc Am.* **88,** 1305-1320 (1977).
2. Artyushkov EV. Stresses in the lithosphere caused by crustal thickness inhomogenities, *J Geophys Res.* **78,** 7675-7708 (1973).
3. Behrmann J, Drozdzewski G, Heinrichs T, Huch M, Meyer W, Oncken O. Crustal-scale balanced cross-sections through the Variscan fold belt, Germany, *Tectonophysics.* **196,** 1-21 (1991).
4. Blundell D, Freeman R, Mueller S. *A continent revealed - the European Geotraverse.* Cambridge University Press, Cambridge, 275 p (1992).
5. Braun J, Beaumont C. Contrasting Styles of Lithospheric Extension: Implications for Differences between the Basin and Range Province and Rifted Continental Margins, *Amer Assoc Petrol Geol and Canadian Geological Foundation Memoir.* **46,** 53-80 (1989).
6. Burg J-P, den Driessche J van, Brun J-P. Syn- to post-thickening extension in the Variscan Belt of Western Europe: modes and structural consequences, *Geologie de la France.* **3,** 33-51 (1994).
7. Byerlee JD Friction of rocks. *Pure Appl Geophys.* **116,** 615-626 (1978).
8. Carter NL, Tsenn MC. Flow properties of the continental lithosphere, *Tectonophysics.* **136,** 27-63 (1987).
9. Cermak V. A geothermal model of the central segment of the European Geotraverse, *Tectonophysics.* **244,** 51-55 (1995).
10. Chopra PN, Paterson MS. The experimental deformation of dunite, *Tectonophysics.* **78,** 453-473 (1981).
11. DEKORP Research Group C. DEKORP 3/MVE-90(West) - preliminary geological interpretation of a deep near-vertical reflection profile between the Rhenish and Bohemian Massifs, *Germany. Z geol Wiss.* **22,** 771-801 (1994).
12. Dewey JF. Extensional collapse of orogens, *Tectonics.* **7,** 1123-1139 (1988).
13. Eisbacher GH, Liischen E, Wickert F. Crustal-Scale Thrusting and Extension in the Hercynian Schwarzwald and Vosges, Central Europe, *Tectonics.* **8,** 1-21 (1989).
14. Franke W. Tectonostratigraphic units in the Variscan belt of central Europe,. *Geol Soc of America, Special Paper.* **230,** 67-90 (1989).
15. Franke W. Phanerozoic structures and events in Central Europe. In: Blundell D, Freeman R, Mueller S (eds) *A continent revealed - The European Geotraverse.* Cambridge University Press, Cambridge, pp 164-180 (1992).
16. Govers R, Wortel MJR Initiation of asymmetric extension in continental lithosphere, *Tectonophysics.* **223,** 75-96 (1993).
17. Handy MR. The solid-state flow of polymineralic rocks. *J Geophys Res.* **95,** 8647-8661 (1990).

18. Henk A. Late orogenic basin evolution in the Variscan Internides: the Saar-Nahe Basin, southwest Germany, *Tectonophysics*. **223,** 273-290 (1990).
19. Henk A Late Variscan exhumation histories of the southern Rhenohercynian Zone and western Mid-German Crystalline Rise: results from thermal modeling, *Geol Rundschau*. **84,** 578-590 (1995).
20. Houseman G, England P. A dynamical model of lithosphere extension and sedimentary basin formation, *J Geophys Res*. **86,** 6115-6132 (1986).
21. Krohe A Structural evolution of intermediate-crustal rocks in a strike-slib and extensional setting (Variscan Odenwald, SW Germany): differential upward transport of metamorphic complexes and changing deformation mechanisms, *Tectonophysics*. **205,** 357-386 (1992).
22. Lorenz V, Nicholls IA. The Permocarboniferous basin and range province of Europe. An application of plate tectonics. In: Falke H (ed) *The continental Permian in Central, West and South Europe*. Reidel, Dordrecht, pp 313-342 (1976).
23. Lorenz V, Nicholls IA. Plate and Intraplate processes of Hercynian Europe during the late Paleozoic, *Tectonophysics*. **107,** 25-56 (1984).
24. Menard G, Molnar P. Collapse of a Hercynian Tibetan Plateau into a late Palaeozoic European Basin and Range province, *Nature*. **334,** 235-237 (1988).
25. Paproth E. Die pal ogeographische Entwicklung Mittel-Europas im Karbon, *Geol Jb Hessen*. **117,** 53-68 (1989).
26. Paterson MS, Luan FC. Quarzite rheology under geological conditions. In: Knipe RJ, Rutter EH (eds) *Deformation Mechanisms, Rheology and Tectonics*. Geol Soc Spec Publ 54, pp 299-307 (1990).
27. Pfiffner AO. Palinspastic reconstruction of the pre-Triassic basement units in the Alps: The Central Alps. In: von Raumer JF, Neubauer F (eds) Pre-Mesozoic Geology in the Alps. Springer, Berlin, pp 29-39 (1993).
28. Ranalli G. *Rheology of the Earth*. Allen & Unwin, London, 2nd ed, 413 p (1995)
29. Shelton G, Tullis J. Experimental flow laws for crustal rocks, *EOS* **62,** 396 (1981)
30. Sonder LJ, England PC, Wernicke BP, Christiansen RL. A physical model for Cenozoic extension of western North America. In: Coward MP, Dewey JF, Hancock PL (eds) *Continental Extensional Tectonics*. Geol Soc Spec Publ, pp 187-201 (1987).
31. Ziegler PA. Geological atlas of Western and Central Europe. 2nd ed Shell *Internationale Petroleum*. Maatschappij, The Hague, 239 pp (1990).
32. Zoth G, Hanel R. Appendix. In: Hanel R, Rybach L, Stegena L (eds). *Handbook of Terrestrial Heat Flow Density Determination*. Kluwer, Dordrecht, pp 449-466 (1988).

Proc. 30th Int'l Geol. Congr., Vol.7, pp. 39-52
Xu Zhiqin, Ren Yufeng and Qiu Xiaoping (Eds.)

Tectonic Characteristics and Orogenic Processes of the Inner Mongolia Orogenic Belt in Eastern China, Asia

YU WANG
Institute of Geology, State Seismological Bureau, Dewai, Beijing 100029, P.R.China

Abstract

Inner Mongolia orogenic belt situated between north China continental block and Mongolia orogenic belt has evolved for a long geological history. Based on the CAAC (continent-arc-arc-continent) collision formed at the late Permian-middle Triassic (200-225Ma) which is later than the end of the subduction in this region, a complicated and duplicated tectonic framework developed with the superimposed structures of fold nappes and thrust nappes. The orogenic vergence is toward south and north continental margins from the core of the orogenic belt, respectively. In the orogenic processes, there were two compression events composed of ductile shearing, folding and thrust systems formation and two times extension or erosion consisted of rift-depression and metamorphic core complex distributed in different positions in the orogenic belt. The superficial tectonic events illustrate interruptions for every tectonic process, but the evolution of the lower crust-upper mantle beneath the orogenic belt maybe remain a continuity during the orogenic processes. The evolutional processes were controlled by the lithospheric dynamics which was not from the far away subduction or collision such as Dabie Shan in central China or Okhotsk in Far East Asia but derived from the tending towards equilibrium of the lower crust-upper mantle themselves.

Keywords: Tectonic Framework, Compression and Extension or Erosion, Orogenic Processes, Inner Mongolia Orogenic Belt, Late Permian, Triassic and Jurassic, Fold nappe and Thrust

INTRODUCTION

Inner Mongolia orogenic belt is situated between Mongolia orogenic belt and north China continental plate in eastern China, Asia (Fig.1).It has evolved a long geological processes. For nearly several decades researches, Chinese and foreign geologists have collected a lot of information on the formation, evolution and some dynamic analysis or hypothesis for this tectonic region [3,6-7,10-11,13-16,18]. Despite these privileged geological and geophysical investigations, details of the evolutional sequences, tectonic framework, mechanisms responsible for Inner Mongolia orogenic belt remain the subject of considerable dispute. Most of former studies dominatedly focused on the subduction of plates, distributions of the ophiolitic suites and "movement classification" by the discordences such as "Hercynian or Indosinian

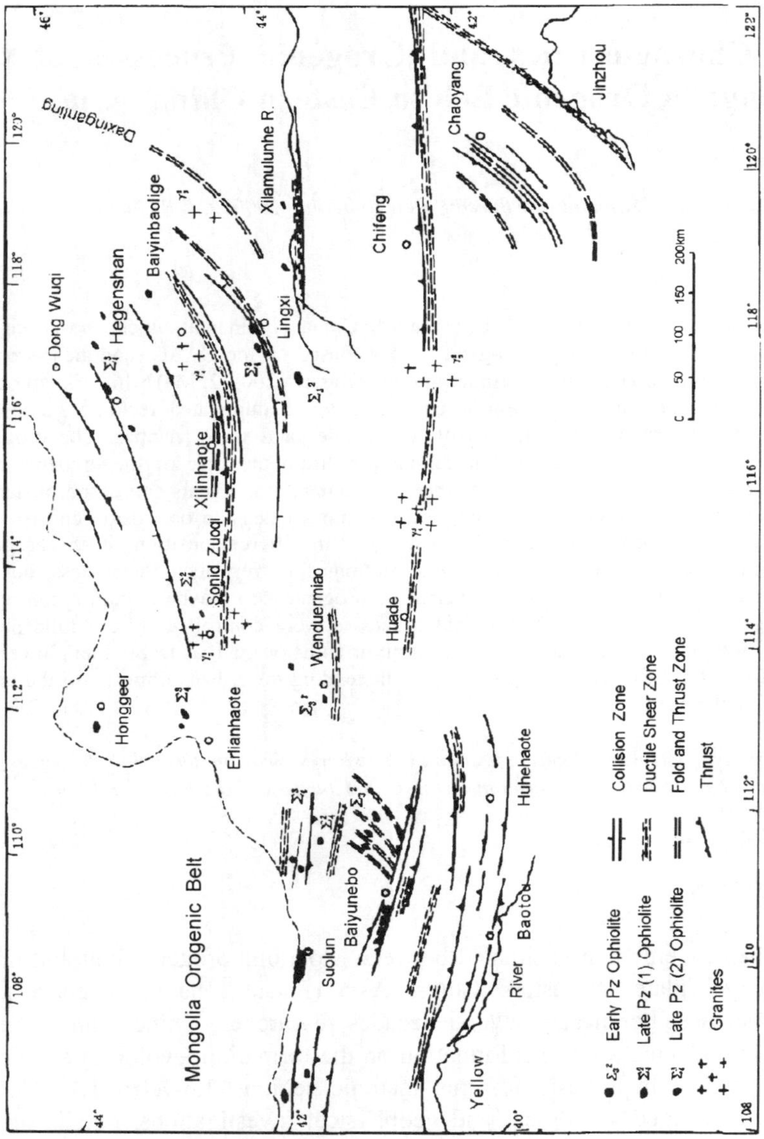

Figure 1. Tectonic map of the Inner Mongolia orogenic belt

Movement". So, several wrong viewpoints exist about its evolutional processes and especially the tectonic events and evolutional sequences after the end of the subduction. In fact, some Chinese geologists have misinterpreted the subduction and collision between Siberia and north China plates.

More over, the evolutional processes during the late Permian-middle Jurassic have not been further through studied. This has directly resulted in a lagged behind of the research levels in continental orogenic belts, relationship between north China and Siberia plates, even though the distinguishing of the basement of some Mesozoic petroliferous basins. And also, in this region, there would be some breakthroughs on geological theories such as the superimposed tectonic sequences, dynamics of the orogenic belt controlled by its lithospheric tending toward equilibrium, the transmission and transformation of the dynamics in the orogenic processes.

TECTONIC BACKGROUND

In this region, prior to the collision in the late Permian, there were a long history evolution. Between Siberia and north China plates, since late Proterozoic, there had evolved the divergences of the continental plates and formation of arc-islands [3,11,14-16]. The arc-islands progressively began to enter collision process, then the arc-islands and the continental margin collided. The paleomagnetic data and polar wandering pathes can support this view [9,17]. But the two continental plates of Siberia and north China had not collided each other, that is to say, no continent-continent collision happened in this region. So this collisional type can be named the continent-arc-arc-continent (simplified as CAAC) collision (Fig.2). The formation and evolution of the Inner Mongolia orogenic belt were based on this tectonic background.

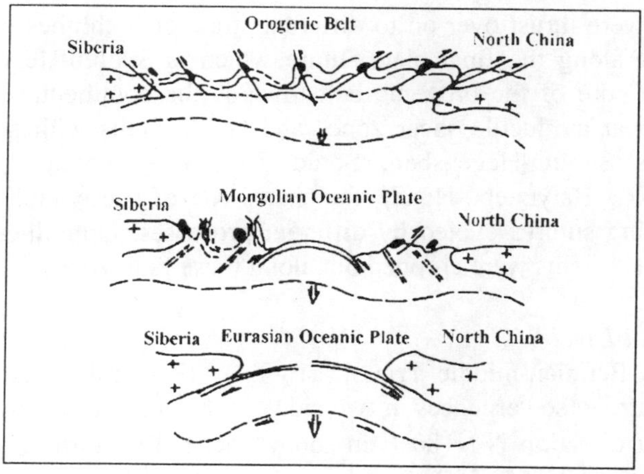

Figure 2. Evolutional background of the Inner Mongolia orogenic belt prior to the late Permian

TECTONIC SETTING

The Inner Mongolia orogenic belt can be divided into several sub-tectonic zones. From north to south, they are core of the orogenic belt named Suolun-Hegenshan tectonic zone, Sonid Zuoqi-Baiyinbaolige zone sited in arc-island regions, Wenduermiao-Xilamulunhe R. tectonic belt near the continental margin and Baiyunebo-Huade tectonic zone in the continental margin. In north side of the orogenic belt, it is adjacent with Mongolia orogenic belt which in fact has similar evolutional processes comparied with the Inner Mongolia orogenic belt (see Fig.1).

The Inner Mongolia orogenic belt changed and was transformed into northeast trend tectonics by the west Pacific tectonic territory at the middle Jurassic [15-17]. The intense deformation and heat-fluid influence developed in the Mesozoic. Just during this period, the ductile shear zones, fold nappes, thrust nappes, magmatic invasion, rift-depression and basins formed in of all the orogenic belt.

In this region, especially in core of the orogenic belt, the lithospheric boot of the orogenic belt has not been found according to the geological transects and geophysical surveys [2,8]. This implies that the statement of the orogenic belt is not same as the characteristics prior to the middle Jurassic.

TECTONIC FRAMEWORK

Through the long evolutional processes, the Inner Mongolia orogenic belt has formed a duplicated and complicated tectonic framework. The subducting process during the late Carboniferous-late Permian (C_2-P_2) caused the thrust sheets of the ophiolitic suites and melange accumulations move towards subducting direction. The ophiolites were thrust over on to different strata or bothlithes in both of south and north sides along the final plate suture which is Suolun-Hegenshan tectonic zone. From the core of the orogenic belt towards the continental margin, several fault zones appear as ductile shear zones and brittle faults with normal or thrust styles, such as Suolun-Hegenshan, Sonid Zuoqi-Baiyinbaolige, Wenduermiao-Xilamulonghe R., Baiyunebo-Huade faults,etc. All of these fault zones are the former subducting sutures evoled by different processes. Ophiolites, ductile shear zones and thrust systems were cropped out along these fault zones.

In the Horizontal Level

During the late Permian-middle Triassic (P_2-T_2), the ductile shear zones formed with almost controverse vergences in two sides of the core of the orogenic belt. The ductile shearing direction is N180°E in south side, but in north side, the shearing direction is towards N355°-5°E. So, these ductile shear zones formed the back-shearing characteristics along the core of the orogenic belt from south to north and north to south, respectively. In this region, there are distributed several ductile shear zones although being broken or uplift on the outcrop by the laterly extension or thrust. The thrust systems have the same moving direction as the ductile shearing. Except in Suolun-Hegenshan fault zone the deformation occurrences illustrate a characteristic of "normal fan type" , others' occurrences express the simple inclined angle, such as dips to north in south side or south in north side of the core of orogenic belt, respectively. Structures of both two sides formed at the same time. This type is different from the backthrust systems like in Dabie Shan orogenic belt or in Gangdese tectonic belt in Himalaya.

Also, in this region, from the core of the orogenic belt toward both sides, the ductile shear zones and thrust systems developed progressively. In the root of the thrusts, the dip angle is 70-80° or much more . Towards south or north, the fault angle is very low, in some positions, less than 5°. In the continental margin, the symmetric thrust type exists in south and north edges of the coal-bearing basins and also the ductile shear zones and thrusts cut and resulted in the uplift and deformation of the lower crust metamorphics, although these thrusts illustrate the thin-skinned brittle thrust features. Until now, the strike-slip faults formed in this orogenic processes have not been found in Inner Mongolia orogenic belt.

On the Vertical Section

Accompanied with the formation of the thrusts, the strata and metamorphics which formed prior to the late Triassic developed folding deformation with recumbent and overturned types. From the core to margins of the orogenic belt, the fold types are different and mainly appears farther away from the core of the orogenic belt, the deformational extent weaker. The deformational sequences are as recumbent fold, towards overturned fold, then symmetric, Jura-type fold finally into the continental block, the fold deformation lost. These fold deformation is composed of the fold nappes accompanied with the ductile shear zones.

The fold nappe type in lower layer was piled up by the later thrust nappe. So in the vertical section, both of these nappes were superimposed (Fig.3). This superimposed tectonic geometry is more complex than the fold and thrust nappes in western Swiss Alps [1].

This vertically superimposed geometry of the orogenic belt is typical in the Inner Mongolia orogenic belt. For this tectonic framework, although the last thrust nappes have not large displacement, the superficial displacement is more than 140km [18]. Some Chinese geologists deliberately misinterpreted the thrust displacement what they realize that the surface dislocation is belong to the final thrust systems in this region. In fact, this is the "looked displacement' of the thrust nappe. Both of the thrust nappes and fold nappes contributed to the final displacement. Fig.4 shows a structural cross section through Inner Mongolia orogenic belt from north to south .

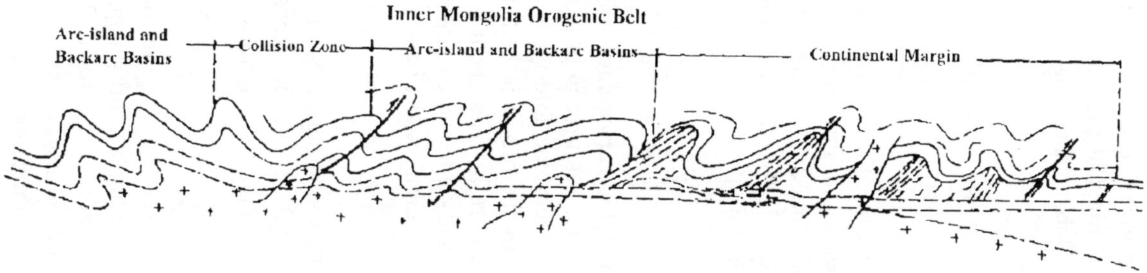

Figure 3. Superimposed model of the tectonic geometry in Inner Mongolia orogenic belt

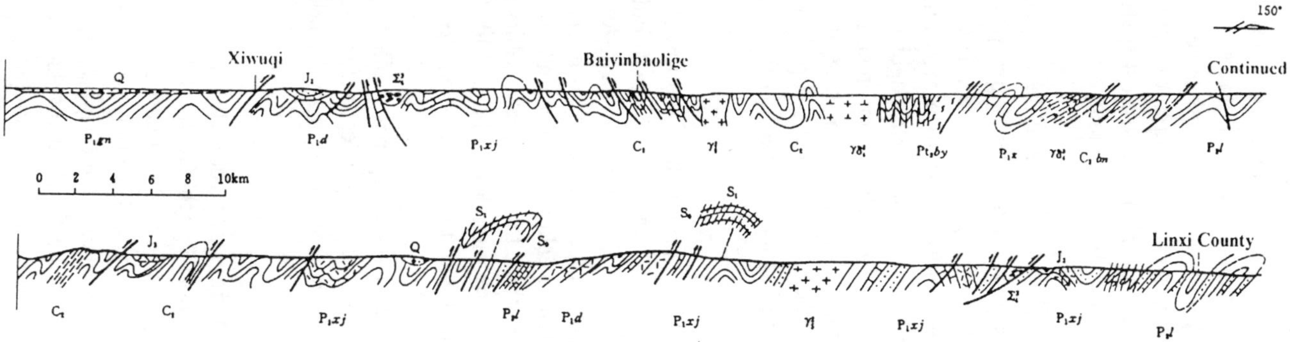

Figure 4. Structural cross section in Inner Mongolia from north to south
Q:Quaternary, J_2:the middle Jurassic, J_1:the upper Jurassic, P_2l:Linxi Formation of the upper Permian, P_1xj:Xiwuqi Formation of the lower Permian, P_1d:Dashizai Formation of the lower Permian, C_2:the middle Carboniferous, C_2bn:Benbatu Formation of the middle Carboniferous, Pt_2by:Baiyunebo Group of the middle Proterozoic, γ_5^2:Mesozoic granites, $\gamma\delta_4^3$:the late Paleozoic granitoid and granites, Σ_4^3:the late Paleozoic ophiolites, S_o:former strata layer, S_1:first stage cleavage. Main legends see Figure 1.

Orogenic Shortening and Vergence

But how to annotate the thrust displacement is a popular problem because of the deformational complex and the erosion in different times. Nevertheless, the orogenic belt shortening can be calculated by the balanced cross section and deformation characteristics. From the core to the continental margin, the shortening is progressively small from 67% to 40%, 30%, then tending to small deformation in intracontinental block.

In view of the vertical and horizontal sections, the vergence of the orogenic belt is toward south and north from the core of the orogenic belt, respectively. This is very similar with the Pyreneans orogenic belt in west Europe [4].

TIME CONSTRAINTS FOR THE TECTONIC EVENTS AND EVOLUTIONAL PROCESSES

In the evolutional processes of an orogenic belt, time sequences are very important for constraints of the geological events, especially the geochronological data. Basically, several tectonic events and magmatic invasion in the Inner Mongolia orogenic belt can be determined. Also, from the different time and events, the relationships and formation sequences among magmatics and tectonics, compression and extension can be built up in this region.

The isotopic data are listed in Table 1. These data illustrate the formation of magmatics, ductile shear zones, basalt volcanics, thrust system and metamorphic core complex.

Table 1. Isotopic data illustrate the magmatics and tectonic deformation in the orogenic belt

Geological Events	Isotopic Data (Ma)	Determining Methods	References
Magamic invasion	200.91+/-4.61,192.27+/-2.57	Rb-Sr (muscovite granites)	[15]
	194,202,225,199,202	K-Ar (granites)	[*]
	186,220,252	U-Pb isotron (granites)	[*]
Ductile shear zones	224.22+/-13.95	$^{40}Ar/^{39}Ar$ plateau age (muscovites in mylonite)	[15,16]
	197.4+/-3.8, 246.3+/-4.7	K-Ar (mylonites)	[15]
Eruption of volcanics	180,182,180,178	K-Ar,U-Pb (Whole rock)	[15]
Extension	157	K-Ar, $^{40}Ar/^{39}Ar$ (mylonites)	[18]

Numbers in brackets indicate reference,[*]unpublished data determined by Weijie Zhang etc. (China University of Geosciences).

Formation of Magmatics

In this region, pre- or post deformation of the ductile shear zones, there were some invasion of magmatics such as granites, mica-granites, muscovite granites and granitoid. They were cut or deformed by the ductile shear zones or folding. Along the several main fault zones, magmatics distributed in every position of the orogenic belt. But for the different components of the granites, the invasive regions were various according to their tectonic background and places far or near the core of the orogenic belt or the central region of the collision zone. Some granites were derived from the upper mantle decomposition and crystallization differentiation were cut by the ductile shear zones which were formed 220-210Ma (see Table 1). Other granites with a lot of muscovites materials formed at 195-200Ma which was later than the formation of the ductile shear zones, but some were formed as same as the shearing formation. These granites were mainly from the lower crust remelting and continental or arc island materials subducting under the mantle and then remelted, invaded up to the middle-upper crust.

Ductile Shear Zones

In the Inner Mongolia orogenic belt, ductile shear zones developed from the core of the orogenic belt to continental margins. At least two times ductile shearing developed in this region. Of the first time, the small scale ductile shearing happened when the ophiolitic thrust sheets were displaced. In the second time, the large scale ductile shearing formed at 210-224Ma (see Table 1). This is same as the formation time of the recumbent folds and syn-tectonic metamorphism all over the orogenic belt. These ductile shear zones cut and deformed Pt-late Permian strata, granites but crossed through by the granites formed at 200Ma. Laterly, there were no any distributions of the ductile shear zones or mylonites even though when the thrust systems took up during the end of early Jurassic-middle Jurassic.

Eruption of Basalt Volcanics

Although there were andesitic series including some basalts formed when the subducting of the oceanic plate and forming of the arc-island, the eruption of the basalt volcanics during the extension is very important and also typical for an orogenic belt. Some orogenic belts, such as Himalaya orogenic belt, have this erupting characteristics during orogenic processes [4]. In Inner Mongolia orogenic belt, the basalts erupted at 180-185Ma, which was after the erosion of the whole orogenic belt and then covered by the early Jurassic coal-bearing sediments.These basalts were derived from the upper mantle and distributed along the north China continental margin, such as in Yanshan area, Liaoxi area etc. This eruption event and later coal-bearing sediments illustrate the extension of the crust and upper mantle layers between the two compressional events which are ductile shear zones and thrust formation.

Formation of Thrusts and Metamorphic Core Complex

In the orogenic belt, the thrusts widely spread. They were covered with the middle Jurassic (J_2) molasse sediments and the middle-upper Jurassic (J_2-J_3) volcanics or pyroclastics. Though there have not got the isotopic data about the thrust time, the strata, volcanics and extensional evidences support that it was formed during the end of early Jurassic-early of middle Jurassic.

The metamorphic core complex is found in the boundary between China and Mongolia. This line is situated in the central of core of Inner Mongolia orogenic belt. The low angle normal fault cuts across the thrust and determined formation time of the mylonites by K-Ar method. The time is 157Ma [18]. At same time, in the core region, the depression and rift-depression developed such as Erlian and Hailaer basins formed their basement. So this process implies the post-orogeny denudation or collapse. After this event, the orogenic belt was influenced by the west Pacific plate subduction under Eurasian continental plate and changed into a new orogenic type [15-16].

EVOLUTIONAL PROCESSES

For the orogenic processes in this region, initiated from the collision of the arc-arc islands, two compressional events and two times extension or erosion were the main episodes. They controlled the evolutional processes and also formed the tectonic framework in this orogenic evolution between Siberia and north China plates. And finally it was evolved into an intracontinental orogenic belt. The end of subduction and initiation of collision, extension and basins' formation, compression development, erosion and low angle extension such as metamorphic core complex which formed same as the northeast tectonic territory development, all of these tectonic events made up of the orogenic processes of Inner Mongolia orogenic belt (Fig.5).

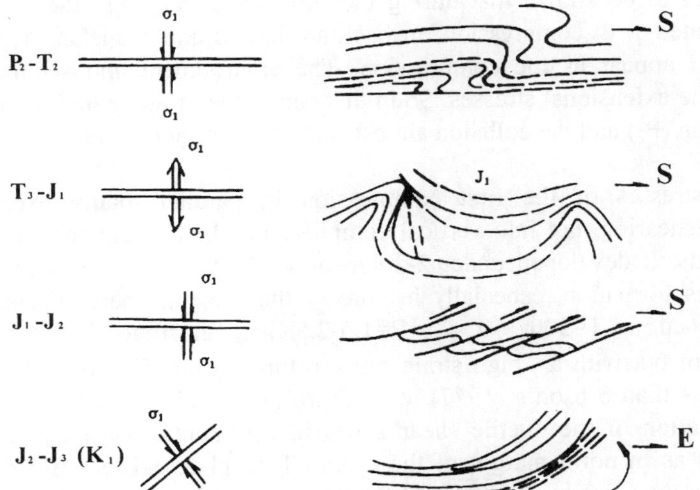

Figure 5. Evolutional processes of the Inner Mongolia orogenic belt

Although in surface of the crust, these tectonic, magmatic and sedimantary events illustrate or imply the interrupted episodes during the orogenic processes, but in the lithospheric layers especially from the lower crust-upper mantle evolution, these express the continuously evolutional processes in this region. It can be seen that from the upper mantle uplift same as the extension of the oceanic plate, then down, towards horizontal feature, then down until the second horizontal characteristics developed, the upper-middle crust evolutional processes were also experienced the subduction and collision, then compression, extension, second time compression and extension or erosion in this orogenic processes. And these processes can make an analogy to the typical orogenic belt in the world. This evolution implies that the orogenic dynamics was not from far away regions, such as some Chinese geologists have guessed these events were controlled by the collision of Okhotsk oceanic plate or from collision of the north China and Yangtze plates in Mesozoic. In fact, these evolutional processes were dominated by the lithospheric tending towards equilibrium or re-balance of the upper mantle, lower crust and some middle crust. All of them had controlled the rigid structures of the upper crust, magmatic invasion and sediments filled into the basins such as the ductile shear zones controlled the folds accompanied with the ductile shearing.

Two Compression Events

Like other typical collision orogenic belts, there are also large thrusts and ductile shear zones in Inner Mongolia orogenic belt. But the special characteristics in this region is that the continuously evolutional processes were interrupted by the extension and relatively period erosion in this region. The first compression was related with the collision of the CAAC, i.e., continent-arc-arc-continent process. Here, it has to be known that during the subducting, although the whole tectonic stresses stated with compression, in back-arc basins and arc-island regions where they didn't appear as the compression. The environment and sediment features illustrate the extensional stresses. So in this paper, the orogeny initially began at the late Permian (P_2) and the collision almost started at the same time.

Tectonic stresses of the first compression in parallel totally exchanged into horizontal shearing, not into vertical compression which could form the thrusts or simple folds. It developed under 2-20km or so depth environment for the ductile shear zones formation, especially in core of the orogenic belts, the ductile shear zones formed at T=300-350 ℃, P=1.5-2.5Kbar, illustrate the shallow layer's environment but with a high strain ratio in this region. This formation depth (2-5km) is less than Sibson's (1977) ideal depth (about 15km) [5,12]. And the same time, formation of the ductile shear zones in continental margin and continental block such as in north margin of the north China plate existed. But these ductile shear zones were produced at very different environment from that of the orogenic belt: T=480-650 ℃, P=4.5-6Kbar, low strain ratios.

This ductile shear zones formed at 211Ma, later than the granites derived from the upper mantle decomposition during subducting at 300-250Ma.The fold deformation was accompanied with the ductile shearing process. Some ductile shear zones and mylonites were also deformed with several recumbent or overturned, inclined folds. Lots of the remelted lower crust materials invaded in the orogenic belt such as mica granites, mainly muscovite granites.The dynamics of this compression event was directly from the subduction-collision of the arc-island and arc-island.

The second compression developed during the end of early Jurassic-prior to middle Jurassic. The intense and wide thrusts in this region dealt with all of the strata and even though some granites came into formation. Certainly, the erosion and extension in the late Triassic-early Jurassic (T_3-J_1) period were very important to this compression because the former event was the basis of the later compression. And the lower Jurassic coal-bearing sediments didn't have a wide distribution and continuous extending in this region. The thrusts cut coal-bearing sediments, but all of these faults, folds, and strata were covered by the middle Jurassic red molasse, volcanics and pyroclastics in all of the orogenic belt even though in north China block. More over, these thrusts were discortantly covered by the middle-upper Jurassic sedimentary basins with northeast axis.

Both of these two time compression mainly resulted in the deep level crust materials uplift in the continental margin. The ductile shear zones of the first compression formed at 25-30km depth with gneiss and amphibole facies, and the second time compression brought the uplift and cropped out of the lower-middle crust materials in the margin of the orogenic belt.

Two Time Extension and Erosion

Between the two compression events, during the late Triassic-early Jurassic (T_3-J_1), there were a widely erosion in this orogenic belt. Mainly, it experienced as a discortance between the lower Jurassic and middle Triassic strata. More ductile shear zones, gravels of mylonites and gneiss coexisted in the upper Triassic-lower Jurassic sediments and in this time, the eroding event existed in all of this region. Although there have not been found the low angle normal fault in this region like the post-orogeny extension, of all folds, ductile shear zones were eroded away in this time. Non-continuous sedimentary basins and depression of coal-bearing sediments of lower Jurassic developed in the orogenic belt. And in the continental margin, the basalt volcanics erupted along E-W strech, which is named Nandaling Formation. The volcanics formed at 185-180Ma was derived from the upper mantle with non-melting of the crust materials. So, this magmatic event controlled by the upper mantle evolution and re-equilibrium or re-balanced of the lithospheric materials and heat texture illustrates the erosion and uplift between two compression events in evolutional processes. After this event, the coal-bearing sediments were built up which implies a warm and wet environment along N40-45° lattitude not only in Inner Mongolia but also spreading in Xinjiang in west China, northeast China and East Asia.

The second extension or erosion was built up after the thrust systems formed in the orogenic belt. In the core of the orogenic belt, on the one hand, the metamorphic core complex cut across the thrust systems formed at 157Ma. On the other hand, the rift-depression basins and depression began to form along the core of the orogenic belt, such as Erlian and Hailarer petroliferous basins. These depressions which were the basement of the Cretaceous petroliferous basins in this belt formed after the middle Jurassic.

This time was also the transfering period for the tectonic territory exchanged from Eurasian territory with E-W trend into the west Pacific plate subducting territory with northeast strike. Of all the erosion caused the middle Jurassic red sediments in the region. Probably, there is a regularity that the first extension of the lithosphere mainly existed in the continental margin. In the second time, the extension dominately developed in the core of the orogenic belt.

But all of the two extension or erosion caused the uplift of the lower crust materials, formation of the eruption of the upper mantle basalt, basins formation with coal-bearing and petroliferous sediments, even though the distributions of the sediments, eruption and the extension of the crust are various and placed in different positions of the orogenic belt.

DISCUSSIONS

In the Inner Mongolia orogenic belt, the compression and extension transformed for at least two times to every event. These tectonic events in this region were not controlled by the outer dynamics. Maybe the lithospheric layers evolution and tending towards equilibrium are the main dynamics. When the subduction developed in Carboniferous-early Permian between Siberia and north China plates, the upper mantle began down from the up state. Then the superficial crust evolved the collision, ductile shearing and folding in the late Permian-middle Triassic, then extension and erosion during the late Triassic-early Jurassic, thrust at the end of early Jurassic-middle Jurassic, final erosion and extension when the middle Jurassic, the upper mantle and part of the lower crust experienced the processes of tending toward down, plat, then down (formed the orogenic belt root) and finally to plat state (re-booting process of the orogenic belt). This is a popular evolutional characteristics perhaps had not been found.

Although the deformation and magmatic events were controlled by the lower crust-upper mantle evolution, the tectonic stresses were transmitted from the core to the margin of the orogenic belt through all of the evolutional processes.

In this orogenic belt, the tectonic framework was built up of the fold nappes (including the ductile shear zones) and thrust systems. They were formed at different time. In some positions, every of them can be investigated lonely. What the displacement of the thrust is not the pure displacement of the final thrust formed.

And also, it has to be known that the formation of thrust systems was not directly related with the ductile shear zones, that is to say, formation of the ductile shear zones was not accompanied with the thrusts, instead of the recumbent and overturned folds. All of the thrusts suggest the thin-skinned tectonic characteristics. The ductile shear zones didn't control the magmatic events.

CONCLUSIONS

According to the above analysis, the conducting remarks can be taken as followed. The orogenic processes of Inner Mongolia orogenic belt was based on a CAAC (continent-arc-arc-continent) collisional evolution between Siberia and north China plates since the subduction end at the late Permian.

The end of Mongolian oceanic plate subduction happened at the late Permian in Suolun-Hegenshan tectonic zone where the ophiolitic suites developed. During the late Permian-middle Triassic, the intense collision event evolved in this zone and also influenced and reacted the north China continental margin.

The tectonic framework of the Inner Mongolia orogenic belt was built up of ductile shear zones, folds, thrust systems, sedimentary basins and lower-middle crust materials. It was composed of two compression events and two times extension or erosion in the orogenic processes from the late Permian to the middle Jurassic. This orogenic geometry appears the feature of "normal fan type" in the horizontal section across the orogenic core from north to south and superimposed section of fold nappe and thrust nappe in vertical section.

Although the superficial tectonic deformation and events illustrate the interruption for every tectonic process, the lower crust -upper mantle beneath the orogenic belt may remain a continuity during the orogeny and post orogeny. The tectonic dynamics transmission had its direction from core to the margin of orogenic belt.

The evolutional processes were controlled by the lithospheric dynamics not from the far away subduction or collision such as Dabie Shan in central China or Okhotsk in Far East Asia but derived from the tending towards equilibrium of the lower crust-upper mantle themselves.

Acknowledgements

I am indebted to Prof. Jishun Ren for his support in fund and guidences during my Ph.D thesis accomplishment in Chinese Academy of Geological Sciences. I should heartily appreciate Prof. Zhiqin Xu for giving me this special chance and Mr.Tiannan Yang for his warm help. I am grateful to Mr. Weijie Zhang etc. for permission to use the isotopic data of granites in Inner Mongolia Autonomous Region and Prof.Wenji Chen, Qi Li for their suggestions. Thanks also should give Miss. Jinping Ji for her help in drawing.

REFERENCES

1. Arthur Escher, Henri Masson and Albrecht Steck. Nappe geometry in the western Swiss Alps, *J.Stru.Geology* **15**,501-509(1993).

2. D.H.Bai, L.Zhang and X.R.Kong. A magnetotelluric study of the Palaeozoic collision zone in the east of Inner Mongolia--Two dimensional modelling: *Acta Geophysica Sinica* **36**, 773-783(1993) (in Chinese with English abstract).

3. R.G.Coleman. Continental growth of northwest China, *Tectonics* **8**, 621-635(1989).

4. M.P.Coward and A.C.Ries (Eds).Collision Tectonics, Published for the Geological Society by Blackwell Scientific Publications, London (1986).

5. R.G.Gibson and D.R.Gray. Ductile-to-brittle transition in shear during thrust sheet emplacement, southern Appalachian thrust belt, *J. Stru. Geol.* **7**, 513-525 (1985).

6. X.Hu, C.S.Xu and S.Y.Niu. Continental margin evolution of the northern margin of North China Plateform during the Early Paleozoic--A series books of plate tectonics of northern China, Beijing University Press (1990) (in Chinese).

7. Inner Mongolia Mineral and Deposition Bureau. Annals of Inner Mongolia geology,Geological Publishing House (1991) (in Chinese).

8. Z.X.Lu and H.K.Xia. Geoscience transect from Dong Ujjimqinqi, Nei Monngol, to Donggou, Liaoning, China, *Acta Geophysica Sinica* **36**, 765-772 (1993).

9. X.H.Ma, L.S.Xing, Z.Y.Yang, S.J.Xu, and J.X.Zhang. Paleomegnetic study in the Ordos Basin since late Paleozoic, *Acta Geophysica Sinica* **36**,68-79 (1993) (in Chinese with English abstract).

10. J.S.Ren, T.Y.Cheng, B.G.Niu, Z.G.Liu and F.R.Liu. Tectonic evolution of continental lithosphere and deposit in eastern China and adjacent areas, Scientific Publishing House (1990) (in Chinese).

11. J.A.Shao. Crustal evolution in central part of the northern margin of Sino-Korea plate--A series books of plate tectonics of northern China, Beijing University Press (1991)(in Chinese).

12. R.H. Sibson. Fault rocks and fault mechanisms, *J.Geol.Soc* **133**,191-213 (1977).

13. K.D.Tang. Tectonic evolution and the deposit rules of the folding belts in northern margin of Sino-Korea plate--A series books of plate tectonics of northern China, Beijing University Press (1992) (in Chinese).

14. Q.Wang, X.Y.Liu and J.Y.Li. Plate tectonics between Sino-Korea and Ankela paleocontinent--A series books of plate tectonics of northern China, Beijing University Press (1991) (in Chinese).

15. Y.Wang.Tectonic evolutional processes of Inner Mongolia-Yanshan orogenic belt in eastern China during the late of late Paleozoic-Mesozoic, Geological Publishing House (1996) (in Chinese with English abstract).

16. Y.Wang. Magmatic thermal events and tectonic evolution of orogenic processes in Inner Mongolia-Yanshan orogenic belt during end of late Paleozoic-Mesozoic, *Geoscience* **10**,66-75 (1996) (in Chinese with English abstract).

17. Y.Zhao et al. The Important exchanges of the tectonic evolution of East Asia, *Geologia Sinica* **29**,105-119 (1994).

18. Y.D.Zheng, S.Z.Wang and Y.F.Wang. An enormous thrust nappe and extensional metamorphic core complex newly discovered in Sino-Mongolian boundary area, *Science in China(series B)* **39**,1145-1152 (1991).

Proc. 30th Int'l Geol. Congr., Vol.7, pp. 53-58
Xu Zhiqin, Ren Yufeng and Qiu Xiaoping (Eds.)

Orogenic Belts of Upper-Amur Region (Using Materials of Morphostructural Studies)

MYANSNIKOV E. A.
Pacifc Institute of Geography, Russian Academy of Sciences, Russia

Abstract

Practical studyes of the orogenic belts of Upper-Amur region (Russian territory of the Upper Amur basin) are quite difficult because of irregularites and insufficient geological studies of some territories. Under these conditions it is better to use the morphostructural analysis which involved: the interpretation of aero-cosmic images and topographic maps, a morphometric analysis of the relief as well as the purpose-oriented analysis of geological and geophysical data. As a result, the orogenic belts of Upper-Amur region (West-Sanovoy, Central-Stanovoy, Umlekano-Ogodzinsky), under study appears, are components of the Amur and Aldan mega Central Type Morphostructure (CTM) of up to 2.5 thousand km in diameter. The spatial location of belts and their geological-geomorphological composition are controlled by arcuate deep-seated faults of mega-CTM. Each orogenic belt under study appears to be a linear elongated CTM of the 20-250 km in diameter. The last display mainly the complex volcanic-plutonic domal-ring structures, whereas small CTM are characterized by a radial-concentric pattern of geolocal and geomorphological conformable complexes. Morphostructural reconstructures indicate that orogenic belts of Upper-Amur region have been mainly in Mesozoic.

keywords: Central Type Morphostructure, arcute fault, linear fault, domal-ring structure, Orogenic belt

STRUCTURE OF OROGENIC BELTS OF UPPER-AMUR REGION

Study of the orogenic belts of Upper-Amur region (Russian territory of the Upper Amur basin) is of great importance, because their origin is related to the tectonic-magmatic activation in the Mesozoic, the main and well-pronounced epoch of endogenous structural formation and ore– mineralization within the limits of the region [1, 2].

Morphostructural studies with the wide use of aero-cosmic images coupled with available geological-geophysical data indicated, that the orogenic belts are well-known volcanic-plutonic structures (West-Stanovoy, Central-Stanovoy, Umlekano-Ogodzinsky) (Fig. 1). They are components of the Amuz and Aldan Central Type Morphostructure (CTM) of up to 2.5 thousand kilometers in diameter. The morphostructures under study have been established by M. G. Zolotov [3], V.V. Solovyov [4], A.P. Kulakov a.o. [5], E.A. Myasnikov [6, 7]. In this case the arcuate deep-seated faults of mega-CTM marginal parts control the spatial location of orogenic volcanic-plutonic belts. Moreover, as a result of large CTM overlap in the Upper-Amur region there is a reasonable chance that the fragmentary manifestation of specific rocks and mineralization of the Umlekano-Ogodzinsky belt occurs within the West-Stanovoy limits, particularly within the zones of deep-seated faults of the north-eastern strike etc.

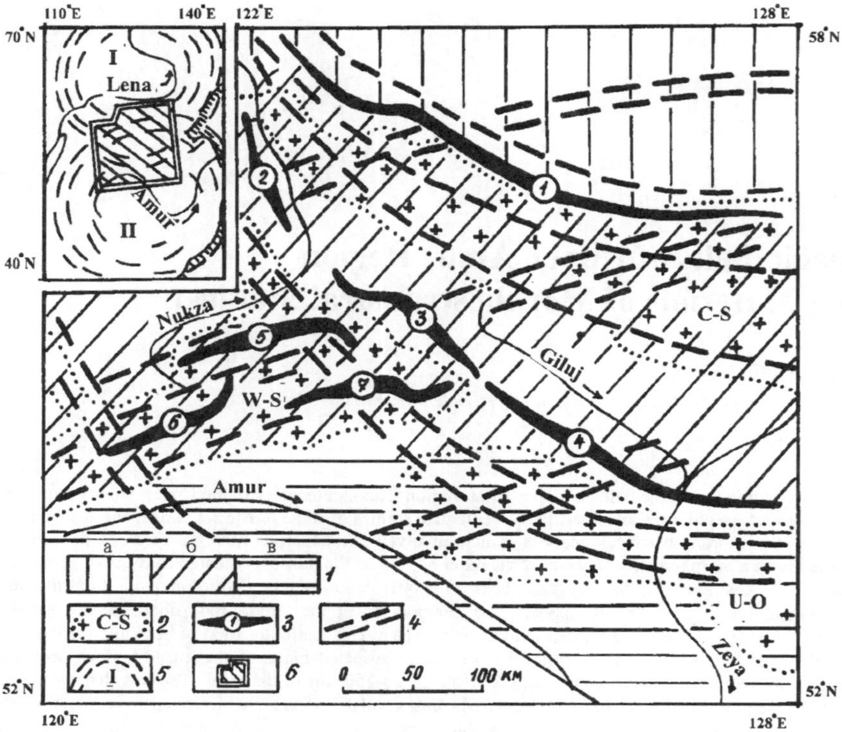

Fig. 1 Orogenic volcanic-plutonic belts and main elements of morphostructural plan of Upper-Amur region
1-crustalline basement blocks (prevail ages in geological indexes): a – Aldan (Ar_{1-2}), б –Stanovoy (Ar_2-Pr_1), B-Amur (Pr_1- Pz_{1-3}); 2 – Mesozoic volcanic-plutonic belts: (C-S) – Central-Stanovoy, (W-S) – West-Stanovoy, (U-O) – Umlekano-Ogodzinsky; 3 – mountain ranges: Stanovik (1), Celbaus (2), Cernysova (3), Tukuringra-Soktachan (4), Dzeltulinsky Stanovik (5), Urusinsky (6), Jankan (7); 4 – arcuate deep-seated faults of mega CTM; 5 – index mega CTM: I – Aldan, II – Amur; Б – boundary of territory morphostructural studies

Each orogenic volcanic-plutonic belt under study appears to be a linear elongated CTM of the third (50-250 km in diameter) and fourth (20-50 km in diameter) orders are controlled by ruptured disturbances of regional morphostructures (mega-CTM and large linear fault zones). Low order CTM are mainly grouped in linear (chain) and isometric (satellite) systems, but in any case they are confined to the zones or nodes of intersection of faults fragments of higher rank morphostructures.

Joint analysis of geological-geomorphic conformable complexes of established morphostructures has indicated that CTM of the third and fourth orders display mainly the complex volcanic-plutonic domal-ring structures, whereas small CTM represent paleovolcanoes and local intrusive domes. CTM of any order are characterized by a radial-concentric pattern of proportionate relief elements of conformable geological complexes and endogenous mineralization.

The effectiveness of our methods and conclusions may be drawn on the example of the West-Stanovoy orogenic volcanic-plutonic belt (Fig. 2). It is located in the southern part of the Aldan-Stanovoy geoblock within the limite of the Mesozoic Yankansky blick-anticline in the near axial part of the Precambrian Mogochin block [8, 9]. In terms of the existing views of the situation the area under study is mainly characterized by linear fault-block structure of the Mesozoic era that is mainly related to endogenous ore-mineralization. In this case the available intrusive and volcanic belts of the Jurassic period are mapped in accordance with established linear blocks, whereas the zones and nodes of intersection of block-separated and cross-cutting faults of diagonal orientation as well as submeridional structures are considered as the main ore-controlling structures.

The conducted morphostructural studies have indicated the existence here of the so-called Yankansky linear-elongated arch, concordant with the block anticline of the same name.

In addition, most of the distinguished linear faults and linear-block morphostructures have been marked on the existing geological-tectonic maps. Nevertheless, it appears that the pattern of the territory under study is mainly identified by a system of 3th, 4th and 5th order CTM. They have conformable sizes, i.e. 50-250, R0-50, 10-20 km in diameter and are in a complex hierarchical coordination in which higher order CTM control the spatial location of lower order CTM. The established CTM become very often the focus of endogenous activity and ore mineralization. We have divided them into three main morphogenetic types, i.e. 1- intrusive-domal; 2-volcanic-domal; 3-volcanic-plutonic domal-ring CTM in the framework of which they are characterized by their specific conformable geological-geomorphic complexes and ore-minernlization.

The typical volcanic-plutonic CTM is the Upper-Khaiktinskaya (50 km in diameter) (Fig. 2). This is a very interesting example when researcheres independent of each other, using different approaches (i.e. the analysis of magmatic formations [10], the interpretation of geophysical anomalies and our morphostructural patterns [6]) drew the same conclusion namely that a magmatic central type structure exists there. Their identified contours were in good agreement on the whole, whereas the conformable structural material, geophysic and morphostructural elements are closely correlates with each other, displaying a radial concentric zonality.

The Upper Khaiktinakaya domal-ring CTM is well-pronounced on cosmic images and topographic maps, particularly its southern part. The northern part is less pronounced, which can evidently be explained by the fact that another CTM is superposed on it (Fig. 2). Gravitation prospecting data provided evidence that a domal form of deep-seated diapir is surely fixed here reflecting thickened area. Its pattern is the result of Mesozoic granitization. An arcuate zone of negative and local positive magnetic anomalies is traced along the periphery of the structure. The CTM focus is located in the intersection node of north-eastern, north-western and sub-latitudinal large transiting faults. The

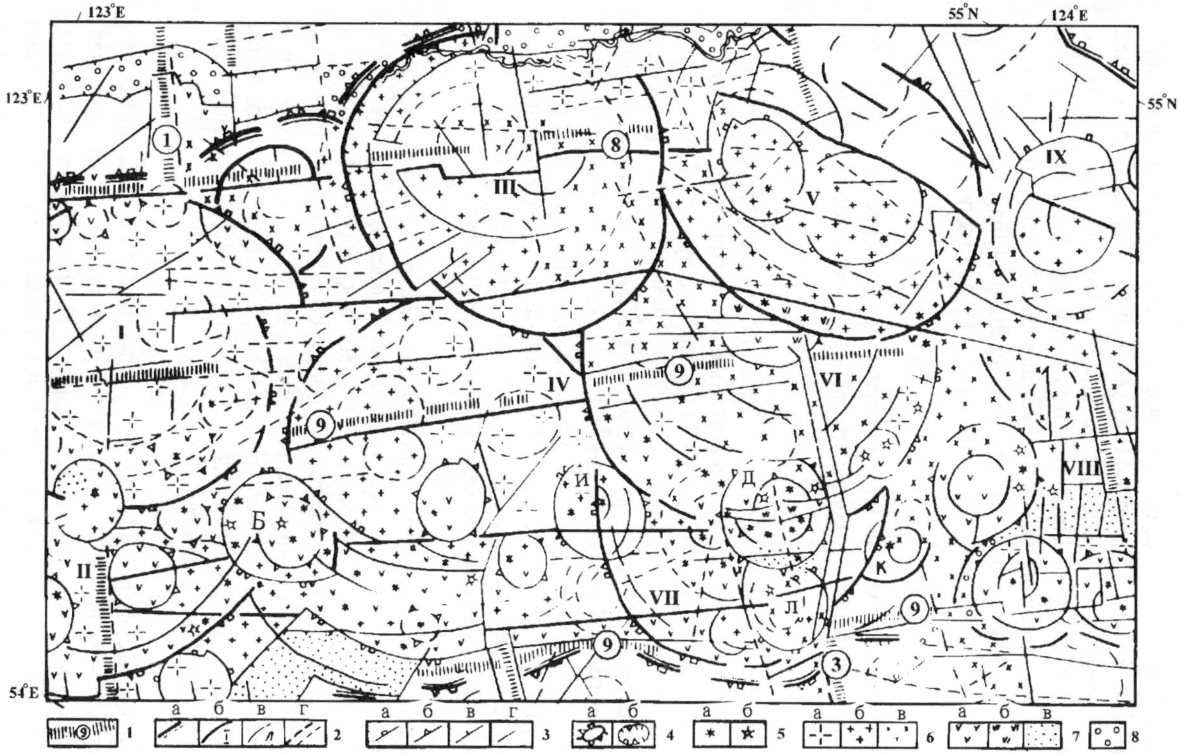

Fig. 2 Structure of the West-Stanovoy orogenic belt using materials of morphostructural studies

1-deep-seated faults of Amur and Aldan mega CTM; 2-arcuate deep-seated faults of CTM: а -(250-500 km in diameter), б -(50-250 km), в -(20- 50 km), г -(10-20 km); 3-principal genetic types of CTM: а -intrusive, б - volcanic, в -dislocative, г -genetic type is unknown, 4-principal morphological types of CTM: а -dome, б -depression; 5-volcanic chimneys: а -by geological date, б -by geophysical date; 6-Mesozoic intrusive rocks and their geological indexes: а -granites ($J_{1\text{-}2}$), б -alkaline granites and syenites ($J_{2\text{-}3}$), в -ganite-porphyry (J_3); 7-Mesozoic volcanic rocks and their geological indexes: a-andesit lavas ($J_{1\text{-}2}$), б -andesit lavas (J_3), в -volcano-sedimentary rocks; 8-Cenozoic sedimentary rocks.

Index and number of main CTM: I: Upper-Urkinskaya, II: Vrkinskaya, III: Upper Nukzinskaya, IV: Amutkazinskaya, V: Oginlynskaya, VI: Upper-Khaiktinskaya, VII: Khaiktinskaya, VIII: Oldoiskaya, IX: Oldoe-Igamskaya, Б : Bolgiktinskaya, и : Ilikanskaya, д : Doeskaya, л : Lazarevskaya identified arcuate and radial faults are evidently related to the establishment of the CTM itself. Inernal and external concentric blocks are well-pronounced high geomorphic marks, arc-shaped drawings of watersheds and hydronet as well as the chains of separate uplifts. These blocks apparently displayed zones of increased penetrability. This is confirmed by the presence of small intrusives, volcanic chimneys and elongated volcanic ares.

The concentric zonality noted above in the distribution of facies of geological rocks and eomorphological indicators as well as the fact that most of the relatively lower temperature ore manifestations are confined to external concenters whereas those of higher temperature are confined to internal concenters indicates their possible paragenetic relationship in the united process of morphostructural development.

Lower order intrusive-domal (10-20 km in diameter) and volcanic-domal MCT (Fig. 2) are widespread within the region. On topographic maps, aerophotos and cosmic images the first ones are characterized by: 1- relatively levelled wide watershed species with weakly expressed watersheds; 2-"II"-shaped transverse profiles of fluvial valleys; 3-relatively gentle slopes; 4-on the whole by little contrast and weakly dissected topography. Geophysical anomalies within the limits of such structures reveal the radial concentric zonality of conformable teotonic disturbances, facies irregularities, hydrothermal and metasomatic changes. The geological pattern of these domes is characterized by concentric facies zonality of the multiphase intrusive massifs composing them, while more acid facies rocks are often located in the CTM central parts. In addition, the thickening of dike sets and manifestations of hydrothermal mineralization was observed mainly along periphery. The developmental stages of intrusive-domal CTM are evidently determined by conformable phases of intrusive activity manifested in intersection nodes of large fracture zones to which they appear to be related genetically.

Volcanic-domal CTM differ considerably from intrusive domes by: 1- relatively higher hypsometric measurements; 2-steep narrow watersheds; 3-"V"-shaped transversal profiles of fluvial valleys; 4-steeply dissected slopes; 5-high contrast and dissected topography. Volcanogenic CTM as well as plutonogenic ones reveal a concentric zonality of their conformable geological-geomorphological complexes and ore mineralization.

SUMMARY

Thus, morphostructural studies have contributed new information on the tectonic pattern and distribution regularities of endogenous ore mineralization within the limits of orogenic belts and geological mapping of the Upper-Amur region. The West-Stanovoy, Central-Stanovoy and Vmlekano-Ogodzinsky orogenic belts are components of the Amur and Aldan mega Central Type Morphostructure (CTM) of up to 2.5 thousand km in diameter. The spatial location of belts and their geological-geomorphological composition are controlled by arcuate deep-seated faults of mega-CTM. Each orogenic belt under study appears to be a linear elongated CTM of the 20-250 km in diameter. The last display mainly the complex volcanic-plutonic domal-ring structures, whereas small CTM are characterized by a radial-concentric pattern of geological and geomorphological conformable complexes. Morphostructural reconstructures indicate that orogenic belts of Upper-Amur region have been mainly in Mesozoic.

REFERENCES

1. *Geological map of the Baikal-Amuz Railway. Scale 1:1500000.* Editor-in-Chief: Krasny L.I. (1977).
2. *Metallogenic map of the Baikal-Amur Railway. Scale 1:1500000.* Editor-in-Chief: Bogdanov Yu.V. (1980).
3. Zolotov M.G. The Pri-Amvr nuclear arched and ring structures, *Tectonics of the Soviet Asian East. Vladivostok:* FEB USSR AS, P. 3-33. (1976)
4. Solovyov V.V. Central Type Morphostructures of the USSR territory using geological-geomorphic data (An Explanatary note to the CTN Map). *L.: VSKGEI,* 110p (1978).
5. Kulakov A.P., Sorokin A.P., Myasnikov E.A. New data on the morphostructure of the BAM eastern zone (using cosmic images), Morphostructural studies in the Far East. *Vladivostok:* FEB USSR AS, P. 48-61 (1983).
6. Myasnikov E.A. On the morphostructure of the Yankanaky uplift (the Upper Amvr region), Morphotectonics of the Far East. *Vladivostok:* FEB USSR AS, P. 81-87 (1981).
7. Myasnikov E.A. Morphostructures of Upper-Amur region (construction, evoluion, practical meanings). *Vladivostok,* 150 p. (1994).
8. Krasny L.I. A regional geology of the Baikal-Amur Railway. *M.: Nedra,* 159 p. (1980)
9. Volsky A. S. The main features of geology and metallogeny of the Urashae Oldoiysky ore region, New data on raw material mineral resources of the BAM central part. *Vladivostok:* FEB USSR AS, 87-94 (1978).
10. Suchov V.I. Volcanogenic formations of the Far Eastern South. *M.: Nedra,* 112 p. (1976)

Proc. 30th Int'l Geol. Congr., Vol.7, pp. 59-74
Xu Zhiqin, Ren Yufeng and Qiu Xiaoping (Eds.)

Early Paleozoic Plate–Tectonic Regime in the Altun Mountains

CUI JUNWEN, TANG ZHEMIN, DENG JINFU, YUE YONGJUN, LI JIXIANG and LAI SHAOCONG
Institute of Geology, Chinese Academy of Geological Sciences, Beijing 100037, China

Abstract

In the Altun Mountains there occur two Early Paleozoic ophiolitic mélange belts: the North Altun ophiolitic mélange belt and the Altun south–margin ophiolitic mélange belt, both of which have very similar rock and mineral assemblages and geochemical features and consist of marine volcanic rocks (ocean ridge basalt, ocean island basalt and island arc basalt) and basic–ultrabasic rocks with strongly deformation and metamorphic fabrics. The main spreading stage of the ocean basin occurred during the Late Cambrian–Early Ordovician. In the Middle–Late Ordovician oceanic crust was subducted northward. In the Silurian the ocean basin ended its evolutionary history. During the Meso–Cenozoic backland thrusting, overthrusting and tectonic spreading in the Qinghai–Tibet region resulted in strong dismembering and displacement of ophiolites. The North Altun ophiolitic mélange belt is the westward extension of the North Qilian ophiolitic mélange belt. The Altun south–margin ophiolitic mélange belt is composed of thrust–superimposed sheets (blocks) of the dismembered Kunlun ocean basin. The initial position of the Early Paleozoic collision between the Yangtze plate and Tarim plate shifts from Central Kunlun to the northern margin of Qaidam. The Early Paleozoic subduction zone on the northern margin of Qaidam resulted from subduction of oceanic crust of the Kunlun Ocean and the Qaidam massif is a thrust–superimposed block formed synchronously with displacement of the Altun south–margin ophiolitic mélange.

Keywords: Altun Mountains, ophiolitic mélange belt, Altun–Qilian ocean basin, Kunlun–Altun–Qilian ocean basin, backland of the Qinghai–Tibet plateau, dismembering, displacement, spreading, thrust–superimposed block (sheet)

INTRODUCTION

The Kunlun Mountains and the Altun and Qilian mountains north of the Kunlun and the Qaidam basin (three mountains and one basin) constitute a region where Early Paleozoic ophiolites are the most widespread and best studied. In the early seventies, following the discovery of ophiolites, high–pressure metamorphic rocks and mélanges in the North Qilian Mountains [17, 21, 19, 27], petrotectonic assemblages related to Early Paleozoic plate motion were found in the Kunlun, South Qilian and Altun mountains as well as on the northern margin of the Qaidam basin [18, 24, 11]. Thus the region has become one of the regions where China's Early Paleozoic plate regimes are best studied. The "three mountains and one basin" is tectonically located in the backland part of the Tethyan–Himalayan orogenic complex [8, 26]. The great dynamic effects produced by the collision of the Indian and Eurasian plates and uplift of the Qinghai–Tibet plateau gave rise to strong deformation, displacement and modification of paleoplate components such as ophiolites and petrotectonic units [7, 26, 9]. Thus people are doubtful about the Early Paleozoic plate regime established according to the present positions of these components.

The Altun Mountains lie in the intermediate area of the backland. There are great differences between the tectonic units on both sides of the mountains. For example, the North Qilian, South Qilian, Qaidam north–margin and Qimamtag ophiolitic mélange zones as the important indications of Early Paleozoic ocean basin spreading end suddenly when they extend westward to the Altun Mountains; although geologists [18, 20, 5, 11] hold that there exists an Early Paleozoic ocean basin in the East Kunlun Mountains, up to now no ophiolites suite as an important indication of ocean basin spreading have been found, while Early Paleozoic island arc magmatic rocks and ophiolite–assemblage rocks are well developed in the northern belt of the West Kunlun [3, 10, 23]. All these undoubtedly show that the Altun Mountains are a key area for the study of the Early Paleozoic plate regime in the backland of the Qinghai–Tibet plateau.

REGIONAL TECTONIC SYSTEMS IN THE ALTUN MOUNTAINS

On the basis of the tectonic evolution and distribution of major regional faults, two regional tectonic systems may be distinguished in the Altun Mountains: the E–W–trending tectonic system and the ENE–trending Altun (active) fault system (Fig. 1).

E–W–trending tectonic system

The E–W–trending tectonic system is composed of the Late Archean–Early Proterozoic Altun north–margin metamorphic complex belt, North Altun ophiolitic mélange belt and Central Altun uplift belt, Late Archean–Early Proterozoic South Altun metamorphic complex belt and Early Paleozoic Altun south–margin ophiolitic mélange belt, which are separated by deep–level ductile shear zones or major regional faults (Fig. 1). In general, the tectonic system is distributed in an E–W direction, and westward it dives beneath the Tarim basin. According to the data of magnetic anomalies it may be inferred that it has a tendency to turn WSW gradually after crossing the Miran–Hongliuyuan fault and that its eastward extension may be cut by the Altun south–margin fault (Fig. 1).

The Altun north–margin metamorphic complex belt occurs as an E–W–trending elongated strip, with its northern side covered by the Tarim basin. It is marked by a high–grade metamorphic series consisting dominantly by gneiss, schist, marble, migmatite and granulite. The metamorphism is of amphibolite–granulite facies [18, 16, 7]. The protoliths are represented by an association of calc–alkaline tonalitic diorite and basaltic volcanic rocks with small amounts of intraplate and oceanic tholeiitic rocks and argillaceous–arenaceous sediments. The belt originated in an environment resembling a modern island arc. Most of the isotopic ages are about 1800–2200 Ma [18, 9], and the oldest isotopic age (U–Pb age) is 2589.37 Ma.

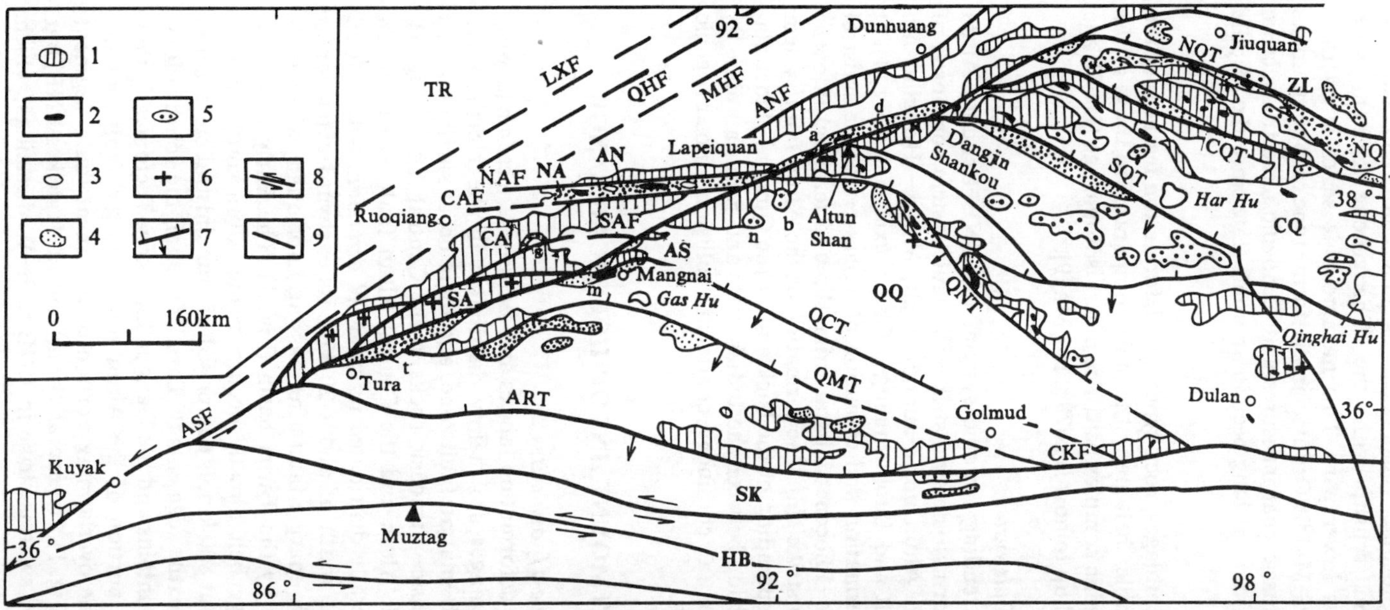

Figure 1. Geological sketch showing distribution of ophiolites in Altun Mountains. 1, Crystalline basement; 2, E. Paleozoic ultrabasic rock; 3, E. Paleozoic basic rock; 4, E.Paleozoic volcanic rock; 5, E. Paleozoic granite; 6, eclogite or blueschist; 7, thrust or ductile nappe shear zone; 8, sinistral strike–slip fault; 9, vertical fault or ductile shear zone. AN, Altun north–margin metamorphic complex belt; NA, N. Altun ophiolitic mélange belt; CA, C. Altun uplift belt; SA, S. Altun metamorphic complex belt; AS, Altun south–margin ophiolitic mélange belt: d, Dangjin Shankou ophiolitic mélange segment; a, south side of the main peak of the Altun Mts. ophiolitic mélange segment; m, Mangnai ophiolitic mélange segment; t, Tura ophiolitic mélange segment; NQ, N. Qilian ophiolitic mélange belt; CQ, C. Qilian uplift belt; QQ, Qilian–Qaidam massif; SK, S. Kunlun terrane; HB, Hoh Xil–Bayan Har terrane; TR, Tarim massif; AL, Alxa massif; ZL, Zoulang backland sag belt. b, Beishan rock body swarm; n, Niubiziliang rock body swarm; s, Suwu Shiji rock body. NAF, N. Altun fault; CAF, C. Altun fault; SAF, S. Altun fault. NQT, N. Qilian thrust; CQT, C. Qilian thrust; SQT, S. Qilian thrust; QNT, Qaidam north margin thrust; QCT, Qaidam central thrust; QMT, Qimantag thrust; ART, Alkatag thrust; CKF, Center Kunlun fault; ASF, Altun south–margin fault; ANF, Altun north–margin fault; MHF, Miran–Hongliuyuan fault; QHF, Qiemo–Heijianshan fault; LXF, Luobuzhuang–Xingxingxia fault.

The Central Altun uplift belt is composed principally of Middle–Upper Proterozoic low–grade metamorphic rocks such as metasandstone, schistose siliceous rocks, phyllite and schist, with the Changcheng System with meta–andesite occurring in its lower part. This metamorphic series is overlain unconformably by some Lower–Middle Ordovician platform–type carbonate rocks and pelite, Lower–Middle Jurassic coal measures and Cenozoic terrestrial clastic rocks, implying that Central Altun has long been in an uplift state since the Middle Proterozoic.

The metamorphic mineral assemblages, rock types and deformation features of the South Altun metamorphic complex belt are in the main comparable to those of the Altun north–margin metamorphic complex belt. The oldest isotopic age (U–Pb age of migmatitic granite) obtained up to now is 2679 ± 142 Ma [9].

ENE–trending Altun (active) fault system

Superimposed on the E–W–trending tectonic system, the ENE–trending Altun fault system is a large–scale normal–sinistral strike–slip fault system composed of the Altun south–margin fault, Altun north–margin fault, Miran– Hongliuyuan fault, Qiemo–Heijianshan fault and Luobuzhuang– Xingxingxie fault (Fig. 1). It has undergone a process of formation and evolution of Triassic–Eocene sinistral strike slip–thrusting, Oligocene–Pliocene sinistral strike slip and terminal Pliocene to the present normal–sinistral strike slip. The formation of the Altun fault system was related to north–directed spreading of the backland of the Qinghai–Tibet plateau resulting from strong collision between the Indian plate and Eurasian plate [6, 9]; so the fault belongs to a continent–continent collision, postorogenic intracontinental transform fault.

EARLY PALEOZOIC ALTUN MOUNTAINS OPHIOLITIC MELANGE

Distribution, composition and age of ophiolites

The Altun ophiolite has strong deformation fabrics and features of tectonic mixing. Early Paleozoic ophiolitic mélanges are mainly distributed in the North Altun Mountains and the Altun south–margin fault zone, and there occur small amounts of Middle–Late Proterozoic basic–ultrabasic rocks in the Central Altun (Fig. 1). Lying between the North fault Altun and the Central Altun fault, North Altun ophiolitic mélange belt is generally distributed in an E–W direction, with a total length of about 400 km and a width of about 20 km. Its eastward extension is directly sited in the Altun south–margin fault after passing Lapeiquan, whereas its westward extension is covered by the Tarim basin (Fig. 1). The Early Paleozoic South Altun ophiolitic mélange belt directly occurs in the Altun south–margin fault. It, with a total length about 800 km, is comprised of four sinistral en–echelon ophiolitic mélange segments, i.e. the segments of Tura, Mangnai, the south side of the main peak of the Altun Mountains and the Dangjin Shankou. Tectonically they are located in the intersection portion of the Altun south–margin fault and the WNW–trending thrust or in the overlapping portion of the linear faults segment and the arcuate fault segment (Fig. 1). In general, the mélange belt strikes ENE, but individual mélange segments or swarms strike in ENE to E–W directions, which are in agreement with the strikes of the thrust.

The ophiolite consists of metamorphosed marine volcanic–sedimentary rocks and basic–ultrabasic rocks. The latters occur in lenticular and elongate forms, with their long axes striking parallel to the trike of regional faults. The marine volcanic–sedimentary series, called previously the Lapeiquan Group (in the North Altun) and the Jiantanshan Group (on the south margin of Altun), is Late Ordovician in age [2, 3]. The Lapeiquan Group may be divided into three formation: clastic rocks and siliceous rocks in the lower part, volcanic rocks in the middle part and clastic rocks and carbonate rocks in the upper part, with a total thickness of nearly 4000 m [3]. Intermediately thick– to thin–bedded siltstone in the lower part of the volcanic formation yields late Late Cambrian trilobites, brachiopods, conodonts and acritarchs [3]; while the clastic formation in the upper part contains the common brachiopod elements in North Qilian, such as Liaguletta sp. and Onyohoplecia sp. [1, 2]. This suggests that the Lapeiquan Group should be late Late Cambrian to Ordovician in age.

Petrological characteristics of ophiolitic melanges

Ophiolitic mélanges in the Altun Mountains have very similar mineral assemblages and petrological–geochemical characteristics [9] and are composed mainly of marine volcanic rocks and basic–ultrabasic rocks with strong deformation and metamorphic fabrics. The major types of ultrabasic rocks are harzburgite, dunite with subordinate serpentinite, lherzolite and pyroxenite. The principal rock–forming minerals are olivine and orthopyroxene and the accessory minerals are magnetite, zircon, chromite, ilmenite and rutile. The olivine content is as high as 65–70% and the orthopyroxene content is generally about 30%. Olivine and pyroxene are commonly serpentinized and there appears serpentinite at places. The main textures are mesh or mesh–rim and blastoporphyritic ones. Basic rocks are dominated by gabbro, which has a medium– and fine–grained gabbroic texture and a massive structure. The principal rock–forming minerals are augite, basic plagioclase and minor quartz. The contents of the three minerals are 50–60%, 35–45% and about 5% respectively. The principal accessory minerals include ilmenite, magnetite, pyrite, zircon, apatite and rutile. Volcanic rocks are mainly basalt, dacite, metatrachyte and rhyolite, which have a pillow structure.

Geochemical characteristics of ophiolitic mélanges

Ultrabasic rocks According to petrochemical data, most samples have lower CaO contents, indicating that the molecular percentage of clinopyroxene is low and that the rock is harzburgite with a small proportion of lherzolite. The rocks belong to low–Al to Al–poor ultrabasic rocks, of which the Al–poor type accounts for the great majority. The rocks have a high H_2O content and has suffered strong alteration. On the Mg / (Fe) versus [Mg+(Fe)] /Si variation diagram, most samples plot in the magnesium field, indicating that ultrabasic rocks consist dominantly of magnesium ones, while ultramagnesium, mafic and ferromagnesium ones are rare (Fig. 2). The total REE content of the ophiolite is very low (about 4.295 ppm), but higher than the average total REE content (0.2965 ppm) of metaperidotite in typical ophiolite. Eu in the ophiolite is slightly depleted with $\delta Eu = 0.92$, whereas metaperidotite in typical ophiolite usually has weak positive Eu anomaly. The REE parameters $(Ce/Yb)_N = 1.004$ and $(La/Yb)_N = 1.17$, showing that there is no marked fractionation of LREE and HREE in the rocks (Table 1).

Gabbro The SiO_2 content in gabbro show narrow variation from 49% to 55%. The ratio of (FeO) / [MgO+FeO] (wt.%) ranges from 0.57 to 0.9. On the diagram of SiO_2 versus $\sum$ FeO / MgO+$\sum$ FeO, data points are dispersed but most of them lie in the field of mafic cumulate and its vicinity (Fig. 3). Gabbro displays notable fractionation of LREE and HREE, with $(La/Yb)_N = 3.95$ and $(Ce/Yb)_N = 3.22$, belonging to the LREE–enriched type. There is marked Eu depletion with $\delta Eu = 0.34$ (Table 2).

Table 1. REE abundance of serpentinite ($\times 10^{-6}$)

La	Ce	Sm	Eu	Tb	Yb	Lu	Gd	$(Ce/Yb)_N$	$(La/Yb)_N$	δEu
0.767	1.79	0.403	0.144	0.103	0.431	0.0653	0.583	1.004	1.17	0.92

Table 2. REE abundance of gabbro ($\times 10^{-6}$)

La	Ce	Nd	Sm	Eu	Gd	Tb	Yb	Lu	$(La/Yb)_N$	$(Ce/Yb)_N$	δEu
36.6	79.8	41.8	9.33	2.24	11.0	1.72	5.99	0.833	3.95	3.22	0.34

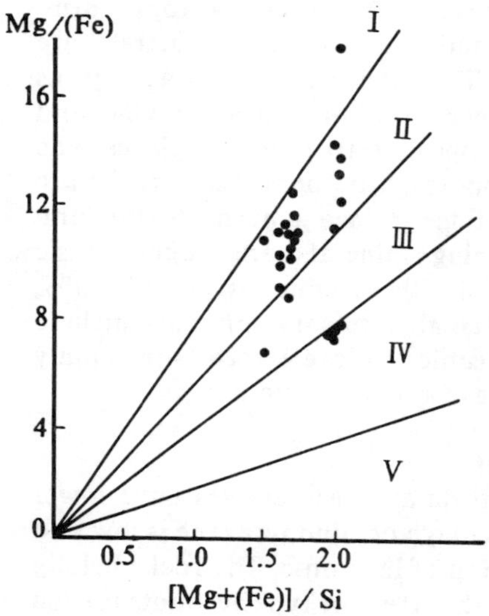

Figure 2. Diagram of Mg / (Fe) versus [Mg+(Fe)] / Si. I, Ultramagnesium; II, Magnesium; III, Mafic; IV, Ferromagnesium; V, Ferruginous.

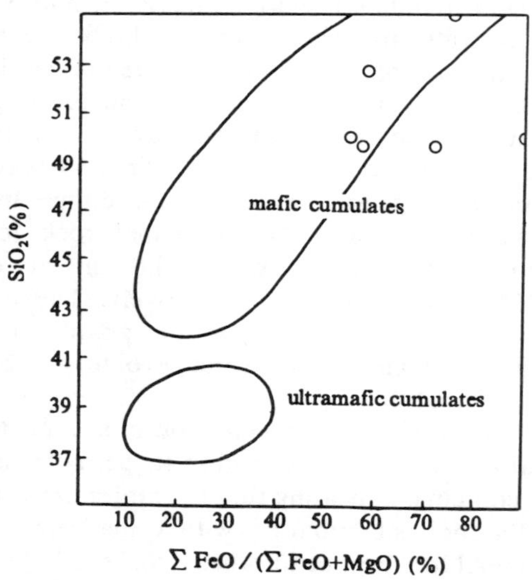

Figure 3. Diagram of SiO_2 versus $\sum$ FeO / $\sum$ FeO+MgO for gabbro (After R.G. Coleman, 1977).

Volcanic rocks On the SiO_2—(K_2O+Na_2O) diagram, the data points of volcanic rocks are concentrated in the fields of basic and acid end–members and on the SiO_2 content frequency diagram, the data points are concentrated in 45–50% and

70–75% intervals; intermediate rocks are lacking [9]. On the SiO_2–(K_2O–Na_2O) diagram, basic volcanic rocks are mostly located in the subalkaline field and only a few samples lie in the alkaline field. On the AFM diagram, two samples of subalkaline basalt display the features of the tholeiite series (Fig. 4). On the Rittmann Suite index (δ) diagram of volcanic rocks (Fig. 5), δ shows a wide variation range. The δ values in the basic and acid rock intervals range from 1.8 to 7 and

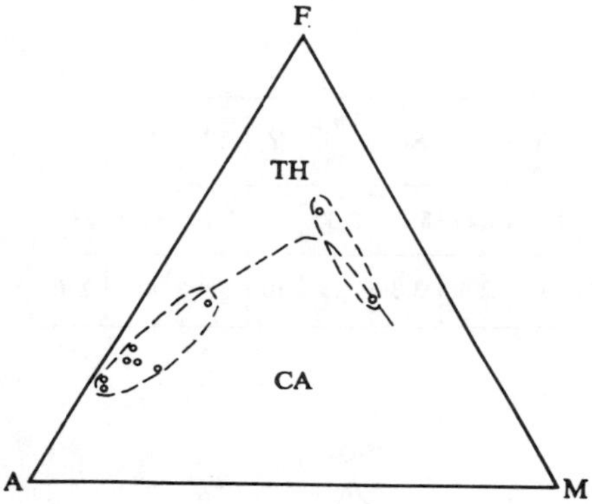

Figure 4. AFM diagram of volcanic rocks. TH, Tholeiite; CA, Calc–alkaline basalt.

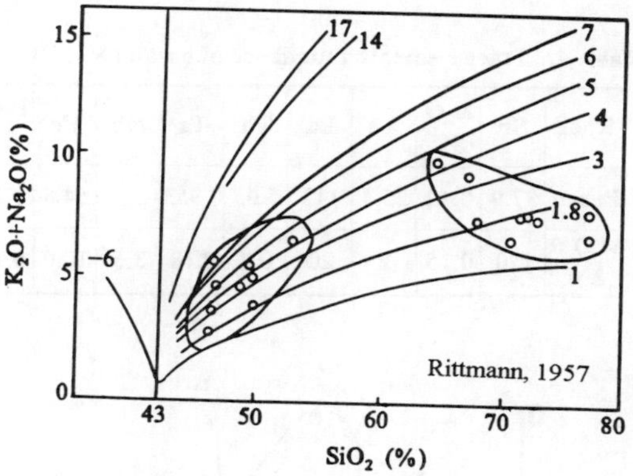

Figure 5. Rittmann suite index and (Na_2O+K_2O)—SiO_2 diagram of volcanic rocks.

Table 3. REE abundance of volcanic rocks ($\times 10^{-6}$)

Rock	La	Ce	Nd	Sm	Eu	Gd	Tb	Yb	Lu	$(La/Yb)_N$	$(Ce/Yb)_N$	δEu
Basalt	24.2	54.8	32.3	8.37	2.51	10.9	1.55	5.79	0.793	2.70	2.29	0.81
Tuff	95.7	189.6	74.95	13.3	2.65	12.4	1.77	6.66	1.04	9.30	6.88	0.61
Tuff	101	188	71.8	11.4	2.34	12.3	1.62	6.02	0.927	10.86	7.55	0.61

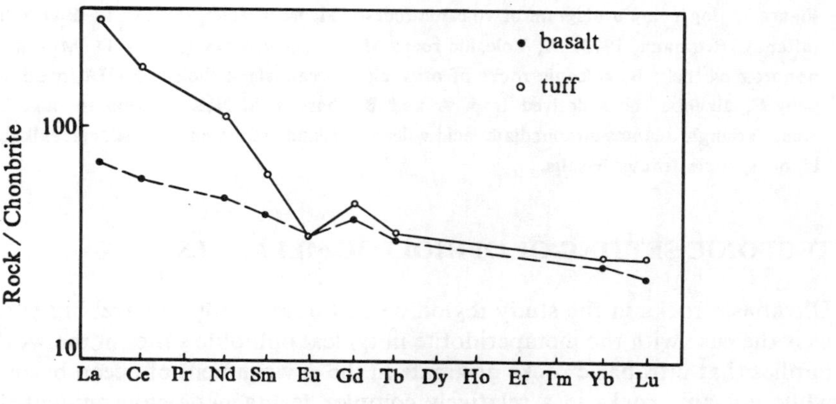

Figure 6. Choudrite–normalized REE distribution patterns of volcanic rocks.

from 1 to 4 respectively. They separately form continuous evolutionary trends, suggesting that they were derived from different magma sources and thus have their own differentiation and evolutionary trends. REE analysis of volcanic rocks (Table 3) indicates that the basalt belongs to the slightly LREE–enriched type. On the REE diagram the REE distribution pattern of the basalt is relatively flat with a small negative slope, and a not–too–distinct depression is produced at Eu. On the other hand, the tuff is of right–inclined LREE–enriched type with relatively strong negative Eu anomaly (Fig. 6).

Table 4. Trace elements abundance of basalt ($\times 10^{-6}$)

Rock	Sr	K_2O %	Rb	Ba	Th	Ta	Nb	Ce	P_2O_5 %	Zr	Hf	Sm	TiO_2 %	Y	Yb	Sc	Cr
Basalt	97.9	0.42	25.4	113	5.65	0.953		54.8	0:21	103	6.21	8.37	2.37		5.79	37.2	21.1
MORB*	120	0.15	2	20	0.2	0.18	3.5	10.0	0.12	90	2.4	3.3	1.5	30	3.4	40	250

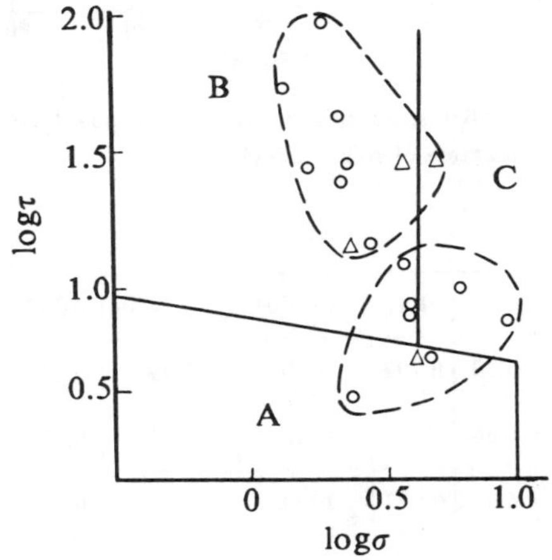

Figure 7. log τ–log σ diagram of volcanic rocks (after A. Rittmann, 1973). A, Volcanic rocks of nonorogenic belt; B, volcanic rocks of orogenic belt; C, alkalibe rocks derived from A and B ereas. Triangle denotes intermediate–acid volcanic rocks; circle denotes basalts.

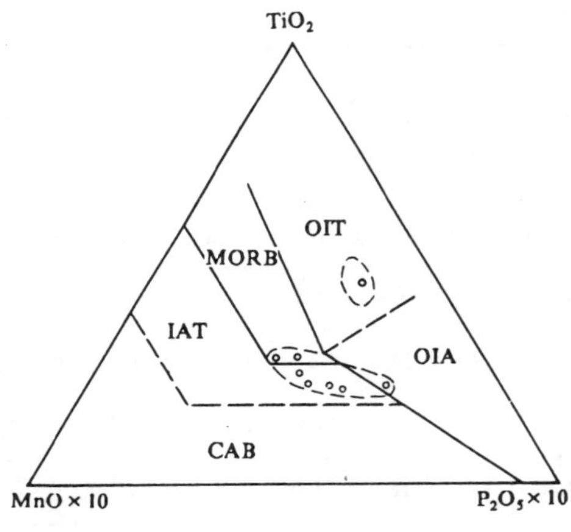

Figure 8. TiO_2–MnO × 10–P_2O_5 × 10 diagram of volcanic rocks (after E.D. Mullen, 1983). OIT, ocean island tholeiites; OIA, ocean island alkalic basalts; MORB, mid–ocean ridge basalts; IAT, island arc tholeiites; CAB, calc–alkaline basalts.

TECTONIC SETTING OF OPHIOLITIC MELANGES

Ultrabasic rocks in the study region consist dominantly of harzburgite and dunite, as is the case with the metaperidotite in typical ophiolite suites in the world [4]. This implies that ultrabasic rocks originate in the environment of ocean basin spreading, while volcanic rocks in a relatively complex tectogenetic environment. On the log τ–log σ diagram (Fig. 7), intermediate–acid volcanic rocks mainly plot in the B field, showing the features of island arc volcanic rocks; while basalt crosses the A, B and C fields [15]. On the TiO_2–MnO–P_2O_5 diagram (Fig. 8), basalt falls in the MORB, IAT, OIT and OIA fields [13]. On the trace–element MORB–normalized pattern diagram of basalt (Table 4, Fig. 9), the pattern exhibits an uplift shape, similar to the trace–element feature of intraplate basalt [14]. Petrochemical characteristics show that the basalt contains both the ocean ridge basalt and ocean–island (seamount) tholeiites related to ocean basin spreading and the island–arc tholeiites related to ocean basin closing and plate subduction.

EARLY PALEOZOIC GRANITOIDS OF THE ALTUN MOUNTAINS

The Early Paleozoic granitic rocks of the Altun Mountains are mainly distributed in the interior of the Altun Mountains and on both sides of the Altun south–margin ophiolitic mélange belt (Fig. 1).

Geological features and ages of typical rock bodies (or rock body swarms)

Beishan rock body swarm at Damenkou of Qaidam   This rock body swarm is located at the intersection portion of the arcuate segment of the Qaidam north–margin sinistral strike slip–thrust fault with the Altun south–margin fault (Fig. 1). It is composed of five small rock bodies with an outcrop area of about 5 km^2 and extends in a nearly E–W direction. The rocks are grayish black and have a fine–grained subhedral texture and a massive structure. The major rock types of the rock bodies are marked by the hornblende diorite–quartz diorite suite. The principal rock–forming minerals are plagioclase (40–50%) and hornblende (40–45%) with subordinate quartz (< 10%). The accessory minerals are apatite and zircon. The whole–rock K–Ar age of the rock bodies is 457 Ma (Late Ordovician).

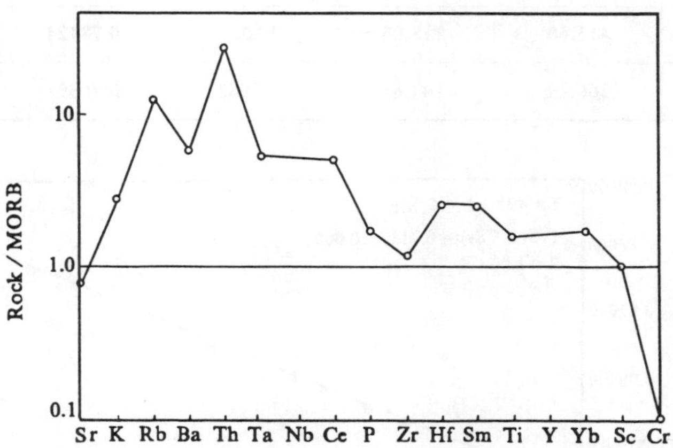

Figure 9. Trace–element MORB–normalized distribution pattern of basalt (after J.A. Pearce, 1973).

Niubiziliang rock body swarm   The rock body swarm is made up of seven small stocks with an outcrop area of 26 km^2. Its country rocks are Early Proterozoic plagioclase gneiss (Fig. 1). It directly occurs in an E–W–trending fault zone. The rock bodies consist dominantly of quartz diorite with subordinate granodiorite. The mineral composition of the former is plagioclase (60–75%), biotite (5–15%), quartz (5–15%) and hornblende (2–25%); the mineral composition of the latter is plagioclase (50–60%), quartz (25–30%), K–feldspar (5–20%) and biotite (2–4%). The accessory minerals are apatite and zircon. The rocks are grayish black and have a medium–grained texture and a massive structure. The rock bodies have a biotite K–Ar age of 503.7 Ma (Early Ordovician).

Suwu Shiji rock body With the carbonate rocks and argillite of the Jixian System as the country rocks, the Suwu Shiji rock body occurs as an elliptical body with a nearly E–W–directed long axis on the northern side of the South Altun ductile shear zone (Fig. 1). The rock body is composed of two different rock types: the core is medium–grained gabbro and the borders are granite, which form a concentric ring, with latter occurring as apophyses penetrating gabbro. Granite is flesh–red in color and has a coarse–grained texture and massive structure, and locally there appears the augen structure. The mineral composition is as follows: plagioclase 20%, microcline 40%, quartz 25–30% and minor biotite. The minerals clearly display the features of plastic deformation. Plagioclase is present in oriented elongate crystals. The rock body has a Rb–Sr isochron age of 491.3 ± 4.6 Ma, formed in the Early Ordovician (Table 5, Fig. 10).

Table 5. Rb–Sr isotopic analytical results of Suwu Shiji rock body

Sample	Rb ($\times 10^{-6}$)	Sr ($\times 10^{-6}$)	$^{87}Rb/^{86}Sr$	$^{87}Sr/^{86}Sr$	$\pm\delta$
AY55–1	461.50	68.308	19.747	0.85008	0.00008
AY57–1	316.76	146.86	6.2473	0.75540	0.0001
AY55–1–2	563.47	125.01	13.116	0.80280	0.00001
AY55–3	415.60	118.08	10.222	0.78424	0.00001
NY57–1–2	366.79	141.61	7.5082	0.76398	0.00007

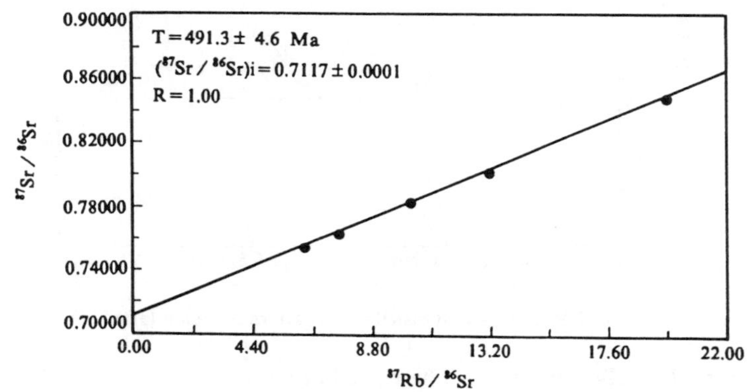

Figure 10. Whole–rock Rb–Sr isochron of the Suwu Shiji rock body.

Geochemical Characteristics of Rocks and Their Tectonic Environment

The Early Paleozoic granitoids in the Altun Mountains form a calc–alkaline magma series with a wide spectrum of composition. Their SiO_2 content is 48–70%. The rocks constitute a complete calc–alkaline magma series of gabbro→ gabbro–diorite→ diorite→ tonalite→ granodiorite (Fig. 11). In this granite series, most rocks are of mantle origin or of crust–mantle mixing origin. The only exception is the Suwu Shiji granite which has high–silica, peraluminous and potassium–rich characters with A / CNK = 1.01–1.07. It originated by partial melting of crustal materials. Gabbro and diorite in the series are characterized by intermediate REE abundances [$\sum$ REE = 139–162 ($\times 10^{-6}$)], more enrichment in LREE than in HREE and absence of Eu anomaly (Fig. 12a), suggesting that the provenance might be mantle. The high REE content, intermeidate–strong negative Eu anomalies (Fig. 12b) and higher Sr initial ratio (I_{Sr} = 0.7117) (Fig. 10) of granites indicate a crustal source. The geochemical features show that the granitoids of the Altun Mountains have the features of active continental–margin island–arc rock assemblages. Their formation was related to subduction of oceanic crust during plate convergence.

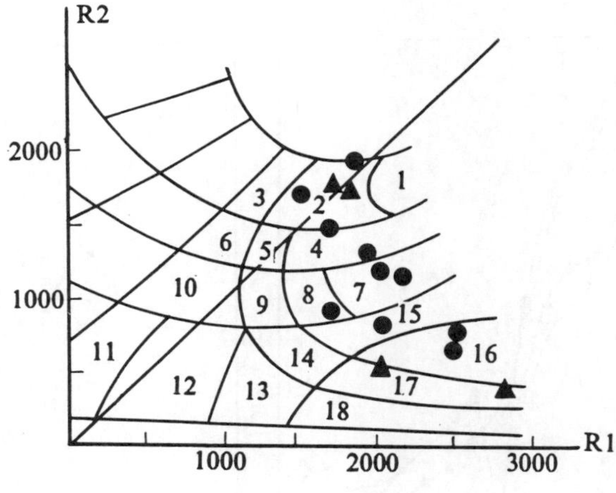

Figure 11. Geochemical classification of Early Paleozoic granitoids R1R2. R1, 4Si–11 (Na+K)–2(Fe+Ti) (ionic number); R2, Al+2Mg+6Ca (ionic number). 1, gabbronorite; 2, olivine gabbro; 3, alkali–gabbro; 4, gabbro; 5, monzogabbro; 6, syenogabbro; 7, diorite; 8, monzodiorite; 9, monzonite; 10, syenodiorite; 11, nepheline syenite; 12, syenite; 13, quartz syenite; 14, quartz monzonite; 15, tonalite; 16, granodiorite; 17, granite; 18, alkali granite. Triangle denotes granitoids north of the Altun south margin fault; circle denotes granitoids south of the Altun south margin fault.

EARLY PALEOZOIC PLATE REGIME IN THE ALTUN MOUNTAINS AND ITS CORRELATION WITH THAT IN ITS NEIGHBORING REGIONS

The Altun Mountains are one of the regions where relatively limited studies of China's paleoplate regimes have been conducted. At present a complete ophiolite sequence has not been set up, but a systematic geochemical study indicates that principal ophiolite elements are developed in the North Altun ophiolitic mélange belt: ultrabasic rocks, ocean ridge basalt and abyssal chert formation. The major rock types of the first element are harzburgite and dunite, which are the same as the major rock types of the rock assemblage of metaperidotite in the world's typical ophiolite suite. They were formed in the tectonic environment of ocean basin spreading. Although active continental– margin calc–alkaline magmatic rocks, high–pressure metamorphic belts and backarc basins related to subduction of oceanic crust have not been found yet in North Altun up to now, the occurrence of island arc basalt in Early Paleozoic marine volcanic rocks may show explicitly that subduction of oceanic crust took place in the North Altun area. The occurrence of nearly upright flow cleavages, mylonitization and tight linear folds in the North Altun ophiolitic mélange [7, 9] indicates that North Altun has undergone strong ductile shear and contraction. It is estimated that the amount of continental crustal shortening caused by folding amounts to 60% [9]. Strong shear and contraction resulted in mixing of basic–ultrabasic rocks with ocean ridge basalt, ocean island basalt and island arc basalt originating in different tectonic environments. It follows that North Altun went through the evolution of oceanic basin spreading→ subduction of oceanic crust→ postorogenic displacement and mixing [7, 8, 9, 26].

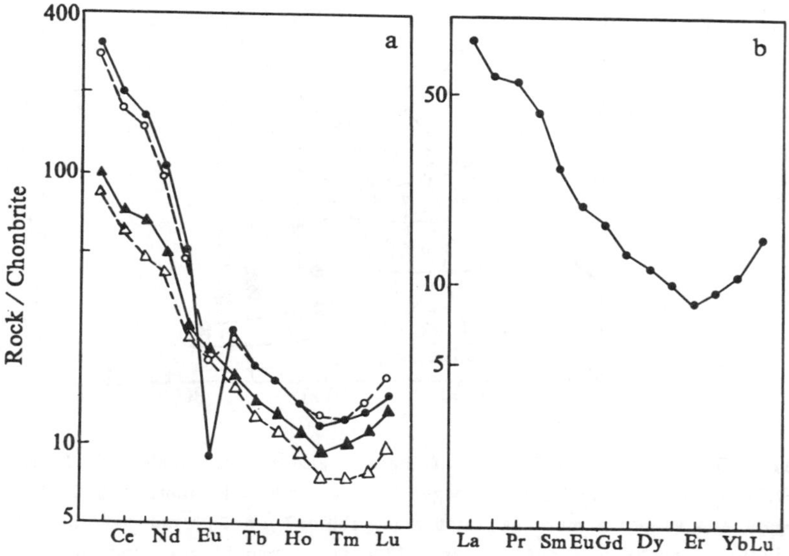

Figure 12. Chondrite–normalized REE distribution patterns of Early Paleozoic granitoids. a. Sample from the Suwu Shiji rock body (triangle denotes gabbro; circle denotes granite); b. Sample from the Beishan rock body at Damenkou of Qaidam.

North Qilian that lies far apart facing North Altun is a region with very typical indications of the paleoplate regime such as Early Paleozoic oceanic crust, ophiolite, trench, island arc, glaucophane schist and mélange [1, 27, 25]. The principal spreading stage of the Qilian ocean basin was the Middle Cambrian–Early Ordovician. The subduction mainly occurred in the Middle Ordovician–Late Silurian and its direction was from south to north [25]. Recent study indicates that there occur not

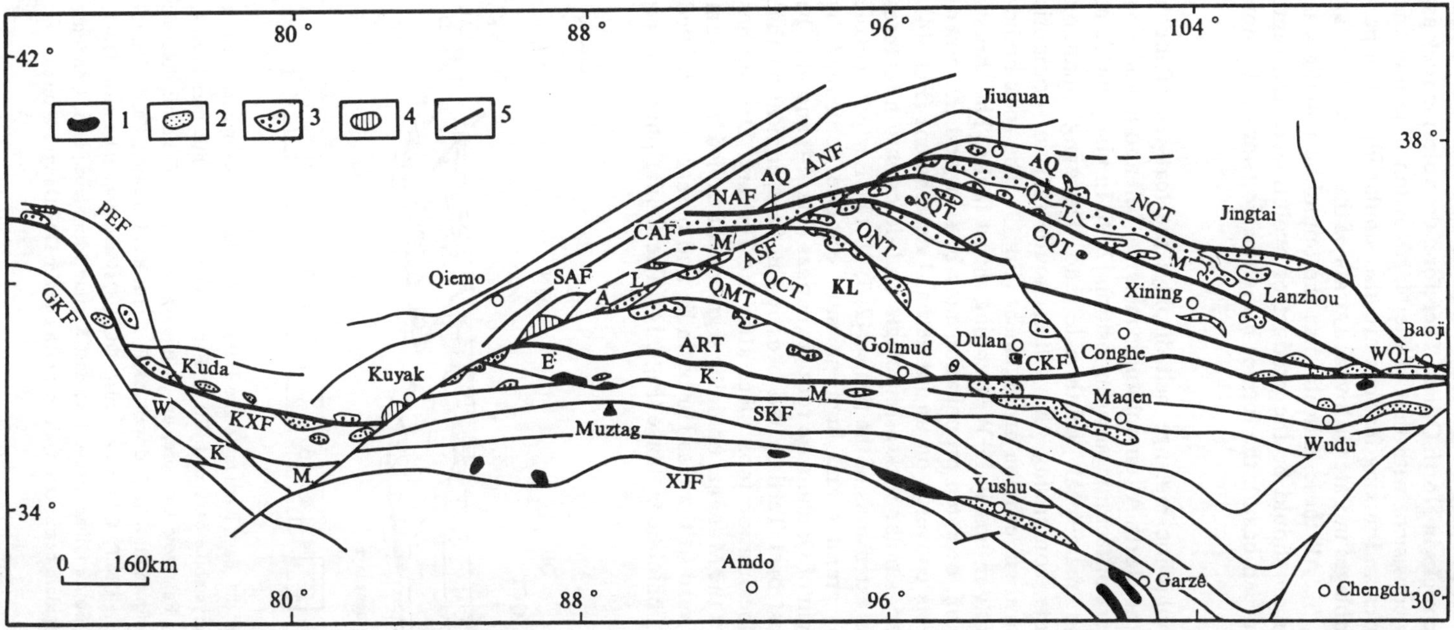

Figure 13. Sketch map showing distribution of ophiolitic mélange and its relation to the faults in the backland of the Qinghai–Tibet plateau. 1, Middle–Late Triassic ophiolitic mélanges; 2, Late Permian–Middle Triassic ophiolitic mélanges; 3, Early Paleozoic ophiolitic mélanges; 4, Middle–Late Proterozoic ophiolitic mélanges; 5, fault. NAF, N. Altun fault; CAT, C. Altun fault; SAF, S. Altun fault; NQT, N. Qilian thrust; CQT, C. Qilian thrust; SQT, S. Qilian thrust; QNT, Qaidam north margin thrust; QCT, Qaidam central thrust; QMT, Qimantag thrust; ART, Alkatag thrust; CKF, C. Kunlun fault; SKF, S. Kunlun fault; XJF, Xijir Ulan–Jinsha fault; ASF, Altun south margin fault; ANF, Altun north margin fault; PEF, Pamir east margin fault; GKF, Karakorum fault; KXF, Kangxiwar fault; AQ, area of ophiolitic mélanges related to Altun–Qilian ocean; KL, area of ophiolitic mélanges related to Kunlun ocean; ALM, Altun Mts.; QLM, Qilian Mts.; WKM, West Kunlun Mts.; EKM, East Kunlun Mts.; WQL, West Qinling.

only typical ocean ridge basalt but also widespread ocean island basalt and island arc basalt in marine volcanic rocks in North Qilian. Therefore, the volcanic rock assemblage in North Qilian is consistent with that of the North Altun ophiolite and their ages, deformation fabrics and mixing features are also similar [9] and in particular the brachiopod assemblages in Late Ordovician strata of the two regions can be entirely correlated [1, 2, 3]. All these indicate that the ophiolitic mélanges in North Altun and North Qilian should be the product of evolution of the same ocean basin (Altun–Qilian ocean basin), the former being the westward extension of the latter (Fig. 13).

The Altun south–margin ophiolitic mélange belt displays the features of much stronger displacement than the North Altun ophiolitic mélange belt and in addition it also shows clearly the nature of tectonic superimposition [7, 9]. In plan, the Tula, Mangnai, Altun Mountain main–peak south–side and Dangjing Shankou ophiolitic mélange segments comprising the Altun south–margin ophiolitic mélange belt display a sinistral en–echelon arrangement (Fig.13). The rock bodies occur as faulted sheets directly in nearly E–W–trending thrust faults. The thrusts generally dip south. In section, overstep propagation may be seen and the downward extensions of the thrusts converge on an intracrustal detachment (Fig. 14). Ophiolitic mélanges with very similar rock assemblages and deformation features are found along the eastward extensions of the thrusts [2, 3, 9]. The Early Paleozoic ophiolitic mélanges, I–type granites and high–pressure eclogite controlled by thrusts on the northern margin of Qaidam are rootless nappes (Fig. 14) and may be directly observed to be thrust over Tertiary red beds in the field, indicating that strong displacement and relocalization took place after the Tertiary. Drilling has demonstrated that southward the Mangnai ophiolite segment controlled by Qaidam central thrust extends downward to the Gas Hu basin [3] and plunges as a faulted sheet beneath the Qimantag ophiolitic mélange (Figs. 14). The similar phenomenon

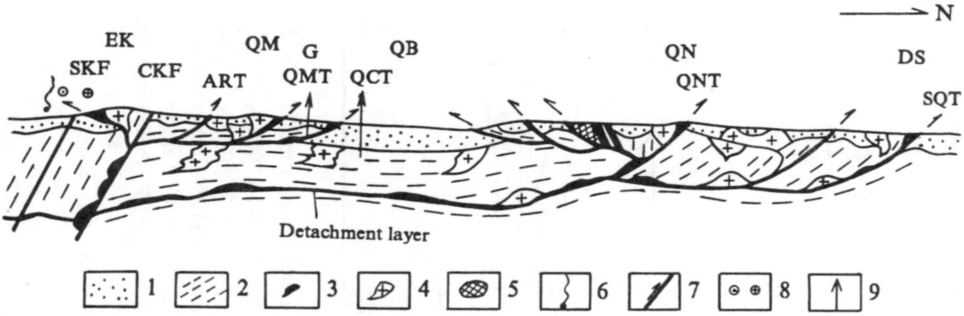

Figure 14. Schematic section from the East Kunlun to the Danghe Nanshan. 1, sedimentary cover (Z–R); 2, strongly flow cleavaged crystalline basement (Pt); 3, Early Paleozoic ophiolitic meranges; 4, Early Paleozoic granitoids; 5, Early Paleozoic high pressure metamorphic rocks; 6, Cenozoic volcanic rocks; 7, thrust; 8, sinistral strike–slip fault; 8, drilled well. SKF, South Kunlun fault; CKF, Central Kunlun fault; ART, Alkatag thrust; QMT, Qimantag thrust; QCT, Qaidam centeral thrust; QNT, Qaidam north margin thrust; SQT, South Qilian thrust. EK, East Kunlun Mountains; QM, Qimantag Mountains; G, Gas Hure Hu; QB, Qaidam basin; QN, Qaidam north margin; DN, Danghe Nanshan.

of sheet overlapping may be also seen in the Qaidam basin north of the line linking Golmud and Dulan [22]. These phenomena suggest that the four ophiolitic mélange segments making up the Altun south–margin ophiolitic mélange belt are connected at depths. They are disrupted and displaced fragments of the Kunlun ocean basin due to strong deep–level shearing and overthrusting with the Central Kunlun thrust as the root zone [6, 8, 9]. The development of the ophiolitic mélange belt, high–pressure eclogite and island arc granite on the northern margin of Qaidam and high–pressure eclogite in the South Altun [12] (Fig. 1) indicates that the Kunlun ocean basin has an evolutionary history of a mature ocean basin as compared with others. The age of middle–high pressure metamorphism with Ar–Ar ages of 453.5 ± 0.8 Ma and 438.69 ± 0.79 Ma on the northern margin of Qaidam [9] should be considered as the time of northward subduction of the Kunlun oceanic crust.

The initial position of the Early Paleozoic subduction zone as an indication for the closing of the Kunlun ocean basin and collision of the Yangtze and Tarim plates is largely along the present Central Kunlun fault zone (Fig. 1). Eastward it goes across the Gonghe basin to link with the Early Paleozoic West Qinling suture zone [24], and westward it extends to West Kunlun and should be located north of the Kangxiwar fault (Fig. 13). The Early Paleozoic calc–alkaline granites (with K–Ar ages of 445–527.6 Ma) at Burumsal, Bilung and Kaxtax of West Kunlun [3, 10] and the granites (Ar–Ar ages of 384–480 Ma) from West Kunlun to Qimantag [23] should be considered as island arc granites related to northward subduction of the Kunlun oceanic crust. The Early Paleozoic West Kunlun suture zone is inferred to be located south of the Burumsal–Kaxtax island arc granite belt. Basic–ultrabasic rocks exposed in the vicinity of Kuyak (Fig. 2) possibly belong to a disrupted fragment of Kunlun Early Paleozoic oceanic crust. According to the data of magnetic anomalies it may be inferred that the eastward and westward extensions of the Early Paleozoic North Altun–North Qilian ophiolitic mélange belt have a tendency to link with the West Qinling and West Kunlun ophiolitic mélange belts (Figs. 1, 13), implying that the North Qilian–North Altun Ocean might have been connected with the Kunlun Ocean in West Qinling and West Kunlun. The bulk of the Kunlun–Altun–Qilian ocean basin should have been along the line of Kunlun–Qinling, and the Central Altun–Central Qilian uplift belts (Fig. 1) was an island chain–like massif in the Kunlun–Altun–Qilian ocean basin [9].

CONCLUSIONS

1. The Early Paleozoic tectonic frameworks of the Altun Mountains and Qilian Mountains may be entirely correlated: lying in the middle was a relatively stable massif (the Altun–Qilian massif), flanked on the south and north by the Kunlun ocean basin and the Altun–Qilian ocean basin. The latter was formed by partial breakup of the continental crust on the southern margin of Tarim resulting from strong spreading of the Kunlun ocean basin between the Yangtze and Tarim plates, belonging to a continental–margin restricted ocean basin. On the other hand, the Kunlun ocean basin was an intercontinental ocean. The Altun–Qilian ocean basin and Kunlun ocean basin have an approximately synchronous formation and evolution process: spreading of ocean basin in the Late Cambrian–Early Ordovician,

north–directed subduction of oceanic crust in the Middle–Late Ordovician, closing of ocean basin in the Silurian, and ending of the history of tectonic evolution characterized by coexistence ocean and land.

2. As the Altun Mountains lie in the backland of the Qinghai–Tibet plateua, the ophiolite, high–pressure metamorphic rocks and island arc granites as important indications of plate motion were subjected to strong displacement and mixing in the process of north–directed thrusting, overthrusting and tectonic spreading of the backland thus most of the rock bodies or rock segments have the nature of nappes and the present location of exposure of the ophiolite belt is not the initial location of plate collision. The North Altun–North Qilian ophiolitic mélange belt possesses the nature of strike slip–superimposition. The North Altun ophiolitic mélange belt was displaced southwestward about 350 km along the strike of the Altun south margin fault relative to the North Qilian ophiolitic mélange belt (Fig. 1). The strong strike slip also led to superimposition and repetition of the North Qilian ophiolitic mélange belt along WNW–trending the thrust (Fig. 1). On the other hand, the Altun south–margin ophiolitic mélange belt, which is made up of four ophiolitic mélange segments of the dismembered Kunlun ocean basin, possesses the nature of thrusting–superimposition. The initial position of collision of the Yangtze and Tarim plates lies along the present Central Kunlun Mountains. Owing to strong north–directed overthrusting and spreading since the Triassic, the Early Paleozoic East Kunlun ophiolitic mélange belt was displaced northeastward about 800 km at most along the strike of the Altun south–margin fault relative to the Early Paleozoic West Kunlun ophiolitic mélange belt (Fig. 13). The bulk of the subduction zone has shifted to the northern margin of Qaidam.

3. The Altun–Qilian massif was an island chain–like block lying between the Kunlun ocean basin and the Altun–Qilian ocean basin or in the Kunlun–Altun–Qilian ocean basin. The long–uplifted Qaidam massif did not exist north of the Kunlun ocean basin, and the so–called Qaidam massif was a thrust–superimposed block displaced synchronously with the Altun south–margin ophiolitic mélange since the Triassic (Fig. 1), which underwent strong downwarping in the Cenozoic.

REFERENCES

1. Bureau of Geology and Mineral Resources of Gansu Province. Regional geology of Gansu province. Geological Publishing House, Beijing (1982).
2. Bureau of Geology and Mineral Resources of Qinghai Province. Regional geology of Qinghai province. Geological Publishing House, Beijing (1982).
3. Bureau of Geology and Mineral Resources of Xinjiang Uygur Autonomous Region. Regional geology of Xinjiang Uygur Autonomous Region. Geological Publishing House, Beijing (1993).
4. R.G. Coleman. Ophiolites, ancient oceanic lithosphere. Springer–Verlag Berlin– Heidelberg, New York (1977).
5. Cui Junwen, Zhu Hong, Wu Changde, Feng Xiaofeng, Tan Zheming and Deng Zongce. Yadong Golmud GGT, deformation and dynamic of the Lithosphere in Qinghai–Xizang (Tibet) plateau. Geological Publishing House, Beijing (1992).
6. Cui Junwen. Deep–level expansion of the lithosphere and uplift of the Qinghai–Tibet plateau—A dis–

Proc. 30th Int'l Geol. Congr., Vol.7, pp. 75-82
Xu Zhiqin, Ren Yufeng and Qiu Xiaoping (Eds.)

A New Tectonic Scheme of the Paleozoides in Mongolia

O. TOMURTOGOO
Institute of Geology, Mongolian Academy f Sciences

The Paleozoides of Mongolia form one of the nearly complete cross-sections of the Inner-Asian Orogen [1, 17] or so-called Altaids [25], which is situated between the Siberian and Sino-Korean cratons. In this paper, the author has analyzed the general tectonic zonation of the Mongolian Paleozoides from the viewpoint of the Concept of terrane collage, which has been developed mainly by American and Canadian researchers [3, 6, 10].

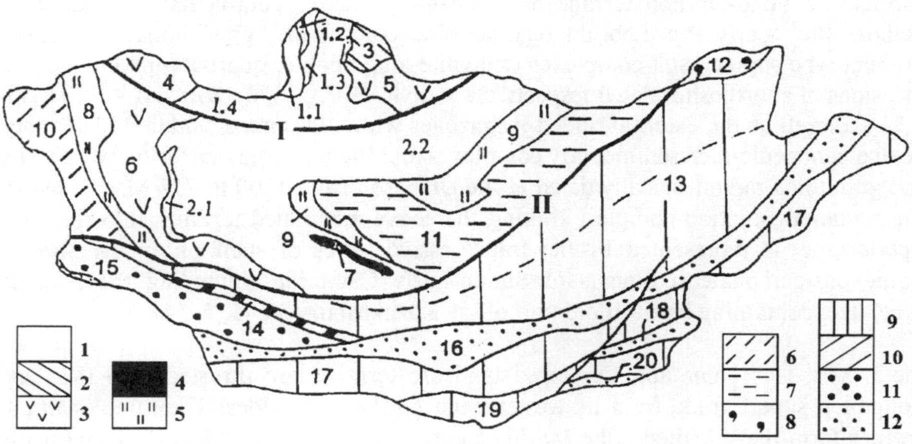

Fig.1. Terrane scheme of the Paleozoides in Mongolia
Northern superblock: 1- superterrane; 2- metamorphic terrane; 3- island arc terranes; 4- terrane of ocean crust; 5-7- turbidite terranes.
Southern superblock: 8- terrane of accretionary prism; 9- terrane of continental margin; 10- terrane of ocean crust; 11- ensialic island arc terrane; 12- ensimatic island arc terrane.
The figures on the scheme are designates the:
Main lineaments: I- Khangay shear-zone and II- Mid-Mongolian Tectonic Line.
Superterrane: 1-2- Tuva-Mongolian (1- northern and 2- soutnern segments), including the Ulaantaigyn (1.1) and Darkhad-Zavkhan-Orkhon (1.3 + 2.2) terranes of continental margin, as well as the Shishged-Dariv terrane of oceanic crust (1.2 + 2.1) and Khugein-Gol terrane of accretionary wedge (1.3).
Composite terranes: 4-6- Lake-Dzhida (terranes: 4- Uvs Nuur, 5- Eg-Uuri, 6- Lake's), 8-9- Khovd-Kharaa (terranes: 8- Khovd, 9- Zag-Kharaa).
Simple terranes: 3- Eastern Khubsugul, 7- Bayankhongor, 10- Mongolian Altay, 11- Khangay-Khentey, 12- Duch-Gol, 13- South Altay-Ereehdavaa (dotted line - Ikhbogd-Kerulen suture), 14- Trans-Altay, 15- Ajbogd, 16- Govian; 17- Govian Tian Shan-Nuketdavaa, 18- Zamin Uud, 19- Toto Shan, 20- Solonkher.

As shows the Terrane Analysis, the folded foundation of the territory of Mongolia are divided into the Northern and Southern superblocks, bordering along newly established Mid-Mongolian Tectonic Line and furthermore are subdivided into numerous multi-aged and multi-typed terranes (Fig. 1).

The Northern superblock of the Mongolian Paleozoides is characterized by distinct mosaic architecture, general nappe structure and large variety of composing terranes, becoming generally younger in the direction to the south.

The most important component for this superblock is the Tuva-Mongolian superterrane which has a structure that is typical to microcontinents [4, 5, 12, 14, 24, 26]. Its basement is very heterogeneous. Recently it was found that in the compound nappe-folded structure there is combination of a four Riphean terrane (from west to east): 1-Ulaantaigyn terrane of passive continental margin, represented by the eastern flank of the known Sangilen Precambrian "Mid-Massif" with the Middle-Upper Riphean carbonates cover often teared away from its metamorphic basement, 2- Shishged-Dariv terrane of ocean crust, forming two allochthon structures with the same type metaophiolite complexes, 3-Khugein-Gol terrane of accretionary wedge, including widespread diabase-greenschist complex with lenses of blueschists [15], and 4- large Darkhad-Zavkhan-Orkhon terrane of Andean-type active continental margin, which includes the "Early Precambrian metamorphic core" with "gray" tonalitic gneisses, different type supracrystal complexes (leptynite-amphibolite, quartzite-marble etc.) and intrusions of anorthosites dated respectively at 2646 (+45), 2370-1830 and ~1700 Ma [4, 9, 31], as well as the external folded megazones with widespread Middle-Late Riphean metavolcanogenic-metasedimentary complexes and the numerous intrusions of different type granitoids radiometrically dated at the Diapason from 1300 to 752 Ma [5, 14, 18]. The postamalgamation complex, linking the above-mentioned terranes into the single superterrane, is represented by the transgressive series of shelf phosphorite-bearing terrigenous-carbonate sediments of Sinian-Early Cambrian (including the Amgaian stage) age, containing basal tilloid unit of the Kazakhstan-type [1, 5, 14].

The Lake's, Uvs Nuur and Eg-Uuri island arc terranes are thrusted onto the Tuva-Mongolian superterrane from its western and northeastern sides. The three island arc terranes form the so-called Lake-Dzhida composite terrane which has a distinct nappe-structure and is included the Troodos-type ophiolites dated at 695+25 Ma (Sm-Nd method), but the main role in its composition are playing the lava of boninite, Ti-high subalkaline and calc-alkaline series, as well as calcarenites, tuffitic turbidites and I-type granitoids radiometrically dated at 480 Ma [7, 8, 12, 14, 16, 17, 26, 27, 30]. The characteristic features of the ophiolites and island arc complexes of the Lake-Dzhida composite terrane clearly indicate its development as a supra-subduction zone.

To the north the Lake-Dzhida composite terrane is emplaced into the East Khubsugul metamorphic terrane which in the regional structure emerges as the south-western end of the Khamardaban cratonic terrane within the southern folded surrounding of the Siberian craton. It consists of the Precambrian-Early Silurian zonal-metamorphic complex with extensive occurrences of granite-dome tectonics [27].

The eastern edge of the Tuva-Mongolian superterrane hosts the elongated but narrow allochthonous Bayankhongor terrane of ocean crust. It is represented by a thick zone of a monomictic serpentinite melange, in large blocks of which an ophiolite section is exposed [7, 14]. The latter includes gabbro dated at 569 Ma (Sm-Nd method), sheeted dike complex and pillow-lava of the E- and T-types MORB with marine beds containing remains of the Early Cambrian fossils [7, 18].

The remaining southern part of the Northern superblock is composed of a system of the Lower and Middle Paleozoic turbidite terranes [1, 11, 12, 16, 17, 29]. Its part, specifically the Khovd-Kharaa composite terrane, in which outcrops mainly the Cambrian-Ordovician volcano-oligomictic turbidites and underlying subduction melange with extensively fractured ophiolite units, is clearly related to the fragment of the accretionary wedge. At that time, the same age Mongolian Altay turbidite terrane with the melange ophiolites and thick subarkosic flysch is most probably a buried mid-oceanic ridge. A separate independent structure in the southern system of turbidite terranes is the Khangay-Khentey terrane which is characterized by thick sections of the extensively dislocated turbidites of the Silurian-Devonian and Lower-Middle Carboniferous. Recently it was revealed that this terrane appears to be a fragment of a small oceanic basin since its Late Silurian-Devonian turbidite Adaatsag-Onon segment is conformably underlain by ophiolite complex with sheeted dike series and pillow-lava of N-type MORB [14].

All terranes in the above-mentioned Northern supreblock covered by the different type overlap complexes, among which main role belongs to the Middle Paleozoic volcanogenic-terrigenous deposits (not including the Khangay-Khentey terrane) and, specially, Late Paleozoic subaeral volcanites associated with different type molasses [2, 11, 12, 16, 17, 26, 29].

The Southern superblock of the Mongolian Paleozoides has a completely different structure compared to the Northern superblock. The former shows a linear structural plan, symmetric character and is formed by a regular alternation (in cross-section) of elongated terranes of continental margin with similarly or narrower-sized terranes composed of paleo-oceanic complexes.

The first class of these structures, consisting from the South Altay-Ereendavaa, Govian Tian Shan-Nuketdavaa and Toto Shan terranes of continental margin, has the Early Caledonian folded basement which, although includes various multi-aged complexes, is rather monotonous and therefore it could be suggested that these terranes are fragments of what used to be a single folded region. The Late Proterozoic essentially carbonate-schist complexes which in places are underlying the Early Precambrian polymetamorphic complexes with "gray" gneisses are predominant in this basement [1, 4, 12]. It was found that the aforementioned complexes are in many features identical to metamorphic rocks of the Zhaertay and Baiyun Obo Groups which fill the "ancient rifts" within the northern margin of the Sino-Korean craton [19, 20]. Higher up in the section in the basement of these terranes lay shelf carbonates of the Sinian-Early Cambrian age with remains of the archaeocyte fauna in the upper section [1, 12]. Only within the Ikhbogd-Kerulen suture in the basement of the South Altay-Ereendavaa terrane, the above shelf carbonates are abruptly replaced by ophiolites of the same age, which are similar to the ophiolite complex in the Ondor Sum suture to the north from the Sino-Korean craton [1, 12, 14, 19, 20]. The Early Caledonian basement of the described continental margin terranes is cross-cut by S-type granitoid intrusions of Middle Cambrian age [1, 12].

The sedimentary cover in the above-mentoined terranes of continental margin is included the Middle Ordovician to Early Carboniferous shelf carbonate-terrigenous complexes, which are in some of the lower horizons are replaced by oligomictic turbidites and in the remaining part of section - by the Silurian-Devonian barrier reefs or Devonian-Early Carboniferous volcanogenic molasse formations [1, 12, 17, 23, 26, 27, 29]. Occasionally, Late Cambrian-Early Ordovician riftogenic volcanogenic-flyshoid complex is present in the composition of the sedimentary cover, and this complex hosts metabasites with petrochemical characteristics of continental tholeiites [23].

Besides, in the terranes of continental margin are represents the different type overlap complexes of Late Paleozoic to Early Triassic age. They are mainly consists of subaeral volcanics and continental or coastal-sea molasses [1, 2, 12].

The second class of the terranes in the Southern superblock occur in the form of paleooceanic sutures [13, 14, 23, 27, 29]. They are characterized by nappe structure and mostly represented by multi-typed volcanogenic-sedimentary complexes which are accompanied by Ligurian-type ophiolites with thin ophicalcite-radiolarite horizon in the base of the upper member of the basaltic pillow-lava. The age interval of these complexes is varying and includes the Silurian-Early Carboniferous in the Trans-Altay terrane of ocean crust, Ajbogd, Govian and Dzamin Uud island arc terranes, and the Middle Carboniferous-Upper Permian in the southmost Solonkher island arc terrane. To the latter similar by age the Duch-Gol terrane of accretionary wedge, that is located on the opposite, northern edge of the superblock. A typical subduction melange of the Devonian-Early Triassic age is exposed in this terrane, which hosts flyshoids that chaotically mixed with greenschist metamorphites and presumably Devonian dismembered ophiolites [17]. It is important to note that the Late Paleozoic complexes of the Solonkher and Duch-Gol terranes differ in their associated faunal complexes: the former commonly hosts a typical Tethys fauna, whereas the latter - Boreal fauna [1, 17, 23].

The formation of terrrane collages in the Mongolian Paleozoides can be interpreted using the following geodynamic model (Fig. 2).

In the Early Proterozoic, all the territory of the Mongolian Paleozoides was entirely a part of the Svecofennian-type mobile belt. Initially, this belt most probably resembled a modern geodynamic system of the Western Pacific Ocean type, but later has concolidated as a result of the Grenvillian tectogenesis and became a part of the supercontinent Pangea I.

During the Riphean-Early Ordovician, the territory of the Northern superblock of the Mongolian Paleozoides found itself within the Paleo-Asian ocean in the form of the

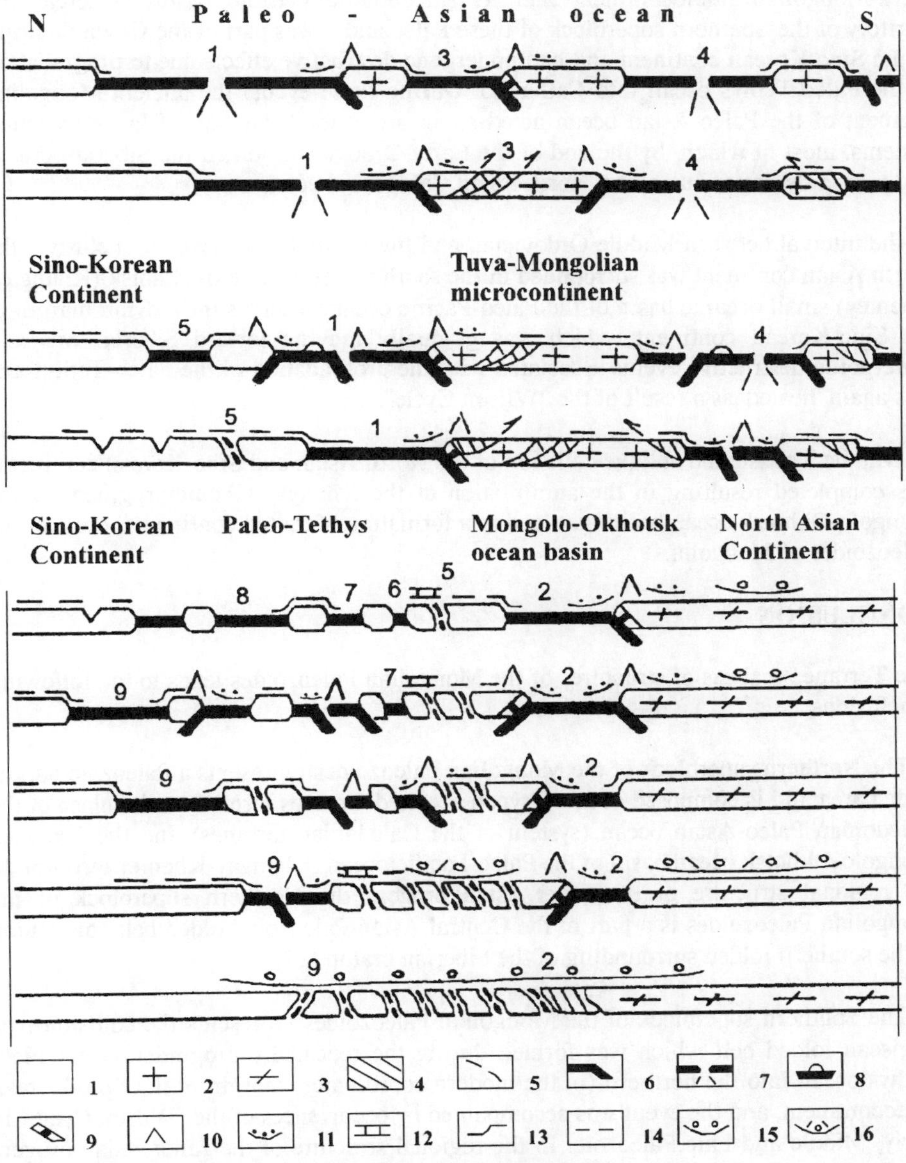

Fig. 2. Plate tectonic model for formation of the Paleozoides in Mongolia
1 - Precambrian Cathaysia; 2 - fragments of Precambrian continental crust of siberian row; 3- Northern Asian Caledonian continent; 4 - accreted terranes; 5 - ophiolitic sutures; 6 - ocean crust; 7- spreading centers; 8- seamounts; 9 - accretionary wedges; 10 - volcanic arcs; 11 -volcanogenic-terrigenous and graywacke series of intra- and marginal basins; 12 - barrier reefs; 13 - oligomictic turbidites of continental slope and rise; 14 - shelf sediments; 15 - molasses; 16 - volcanogenic-molasse complexes.
The arabian figures on scheme are marks the oceanic basins and forming on their place the accreted terranes and sutures: 1, 3 and 4 – oceanic basins of the Paleo-Asian ocean (1 –Altay-Khentey, 3 – Shishged and 4 – Dzhida); 2 – Khangay-Khentey oceanic basin of the Paleo-Pacific ocean; 5-6 – destructive troughs with ocean-type crust (5 – Ikh Bogd-Kherlen and 6 -Govi Altay) and oceanic basins (7 – Trans-Altay, 8 – South Mongolian and 9- Inner Mongolian) of the Paleo-Tethys ocean.

Tuva-Mongolian microcontinent and its surrounding oceanic basins, whereas the territory of the Southern superblock of these Paleozoides was part of the Govian margin of the Sino-Korean continent and had undergone destructive effects due to progradation of the Paleo-Tethys ocean into Cathaysia. During these events the ancient-Mongolian segment of the Paleo-Asian ocean acted as an arena for formation of large accretion systems, most of which, by the end of the Early Ordovician, joined the Siberian craton and turned the latter into a vast North Asian Caledonian continent.

In the interval between Middle Ordovician and the beginning of Triassic inclusive, the North Asian continent was surrounded in the south by the Mongolo-Okhotsk (Khangay-Khentey) small oceanic basin of the Paleo-Pacific ocean, whereas the Govian margin of the Sino-Korean continent, which was gradually moving toward Siberia, was still undergoing destructive events associated with the progradation of the Paleo-Tethys and was again 'healed' as a result of the "Wilson Cycle".

By Middle Triassic the oblique collision of the North Asian and Sino-Korean continents was completed resulting in the annihilation of the Khangay-Khentey segment of the Mongolo-Okhotsk ocean basin that lead to a formation of united continental crust in the Paleozoides of Mongolia.

CONCLUSION

The Terrane Analysis of structures of the Mongolian Paleozoides leads to the following conclusions.

1. The Northern superblock of the Mongolian Paleozoides represents a Paleozoic nappe-folded area and is composed of multi-typed accreted terranes formed on the place of the Caledonian Paleo-Asian ocean (system of the Caledonian terranes) and the Variscan Mongolo-Okhotsk ocean basin of the Paleo-Pacific ocean (Khangay-Khentey terrane). In the regional structure of the Inner-Asian orogen, the Northern superblock of the Mongolian Paleozoides is a part of the Central Asian Paleozoic Folded belt concerning to the southern folded surrounding of the Siberian craton [17].

2. The Southern superblock of the Mongolian Paleozoides represents the collision-type Variscan folded belt which was formed during the repeatedly progradation of Paleo-Tethys ocean into the northern (in the modern coordinates) margin of the Sino-Korean paleocontinent, and the event was accompanied by occurrences of the "Wilson Cycle" in Early, Middle and Late Paleozoic. In the regional structure of the Inner Asian orogen, the Southern superblock of the Mongolian Paleozoides belongs to the Irtysh-Khingan Folded belt in the northern folded surrounding of the Sino-Korean craton [28].

3. The lineament boundary which in Mongolia divides the aforementioned accretionary and collisional folded belts of the Inner-Asian orogen and which is called the Mid-Mongolian Tectonic Line, is a western continuation of the known Mongolo-Okhotsk Lineament [17] into the depths of Central Asia and is depicts the trace of the oblique collision of the North Asian and Sino-Korean paleocontinents. Consequently, the Mid-Mongolian Tectonic Line represents the segment of a fundamental suture of ancient

continental massifs of the Laurasian and Gondwanian rows.

REFERENCES

1. Yanshin A.L., ed. *Tectonics of the Mongolian People's Republic. Moscow: Nauka.* 284 pp (in Russian) (1974).
2. Mossakovskii A.A., Tomurtogoo O. *The Upper Paleozoic of Mongolia. Moscow: Nauka.*126 pp (in Russian) (1976).
3. Coney P.J., Jones D.L., Monger J.W.H. Cordilleran suspect terranes, *Nature (London),* **288,** 329-333 (1980).
4. Mitrofanov F.P., Kozakov I.K., Palei I.P. *Precambrian of the West Mongolia and South Tuva. Leningrad: Nauka.* 156 pp (in Russian) (1981).
5. Iiyin A.V. *Geological development of the South Siberia and Mongolia in the Late Precambrian-Cambrian. Moscow: Nauka.* 115 pp (in Russian) (1982).
6. Willams H., Hatcher R.D. Suspect terranes and accretionary history of the Appalachian orogen, *Geology,* **10,** 530-536 (1982).
7. Zonenshain L.P., Kuzmin M.l., Tomurtogoo O., Kopteva V.V. Ophiolites of the West Mongolia. In: *Riphean-Lower Paleozoic ophiolites of North Eurasia. Novosibirsk: Nauka.* pp. 7-18 (1985).
8. Gordienko I.V. *Paleozoic magmatism and geodynamics of the Central-Asian folded belt. Moscow: Nauka.* 238 pp (in Russian) (1987).
9. Sukhanov M.K., Troitskhii V.A., Bayarbileg L. Evidence of Precambrian age for the anorthosites in Mongolia, *Dokl. AN SSSR.* **298: 4,** 952-955 (in Russian) (1988)
10. Coney P.J. Structural aspects of suspect terranes and accretionary tectonics in Western North America, *Structural Geology.* **11:1/2,** 107-125 (1989)
11. Dergunov A.B. *Caledonides in Central Asia. Moscow: Nauka.* 192 pp (in Russian) (1989)
12. Yanshin A.L., ed. *Map of geological formations of the Mongolian People's Republic. Sc. 1:1,500,000. Novosibirsk* (in Russian) (1989).
13. Ruzhentsev S.V., Pospelov I.I., Badarch G. Tectonics of the Indosinides in Mongolia, *J. Geotektonika,* **6,** 13-27 (in Russian) (1989).
14. Tomurtogoo O. Ophiolites and formation of folded belts in Mongolia, *Doctor thesis. Moscow: Geological Institute of the USSR Acad. Sci.* 59 pp (in Russian) (1989).
15. Sklyarov E.V., Postnikov A.A. The Khugein blueschist belt in North Mongolia, *Dokl. AN SSSR.* **315 (4):** 950-954 (in Russian) (1990).
16. Volochkovich K.L., Leont'ev A.N. The problems of geology of the Mongolian Altay in the light of new data. In: *Evolution of geological processes and metallogeny in Mongolia. Moscow: Nauka.* pp.122-139 (1990).
17. Zonenshain L.P., Kuzmin M.l., Natapov L.M. Geology of the USSR: A Plate-Tectonic Synthesis, *Geodyn. Ser., 21. Washington DC: Am. Geophys. Union; Boulder, CO: Geol. Soc. Am.* 242 pp (1990).
18. Kepezhinskas P.K., Kepezhinskas K.B., Pukhtel I.S. 1991. Lower Paleozoic oceanic crust in Mongolian Caledonides: Sm-Nd isotope and trace element data, *Geophysical research letters.* **18:7,** 1301-1304
19. Ishii Ken-ichi, Liu Xueya, Ichikawa Koichiro, Huang Benhong. *Pre-Jurassic Geology of Inner Mongolia, China. Report of China-Japan Cooperative Research Group, 1987-1989. Osaka.* 236 pp (1991).
20. Wang Quan, Liu Xueya, Li Jinyi. *Plate tectonics between Cathaysia and Angaraland in China.* Beijing. 151 pp (1991).
21. Bibikova Ye.V., Kirnozova T.l., Kozakov I.K., et al.. Polymetamorphic complexes of the southern slope of the Mongolian and Govian Altay, *J. Geotektonika* **2,** 104-112 (in Russian) (1992)
22. Ruzhentsev S.V., Pospelov I.I., Badarch G.. Tectonics of the Barunkhuray Depression (Mongolia), *J. Geotektonika* **1,** 95-110 (in Russian) (1992)
23. Ruzhentsev S.V., Pospelov I.I. The South-Mongolian Variscan fold system, *J.Geotektonika.* **5,** 45-62 (in Russian) (1992).
24. Makarychev G.l. 1993. Tectonics of Tuva-Mongolian massif, *Bull. MOIP.* **68:5,** 36-48 (1993)
25. Sengor A.M.C., Natal'in B.A., Burtman V.S.. Evolution of the Altaid tectonic collage and Paleozoic crustal growth in Eurasia, *Nature.* **364,** 299-307 (1993)
26. Zorin Y.A., Belichenko V.G., Turutanov E.Kh., et al.. The South Siberia-Central Mongolia transect, *Tectonophysics* **225,** 361-368 (1993).
27. Belichenko V.G., Sklyarov E.V., Dobretsov N.L., Tomurtogoo O. Geodynamic map of Paleoasian ocean: eastern segment, *J. Russian Geology and Geophysics.* **35:7-8,** 23-32 (1994).
28. Didenko A.N., Mossakovskii A.A., Pecherskii D.M., et al.. Geodynamics of the Central-Asian Paleozoic Oceans. *J. Russian Geology and Geophysics.* **35:7-8,** 48-61 (1994).
29. Zorin Yu.A., Belichenko V.G., Turutanov E.Kh., et al.. Baikal-Mongolia transect, *J. Russian Geology and Geophysics.* **35:7-8,** 78-92 (1994).

30. Konnikov E.G., Gibsher A.S., Izokh A.E., Sklyarov E.V., Khain E.V. Late-Proterozoic evolution of the northern segment of the Paleoasian ocean: new radiological, geological and geochemical data, *J. Russian Geology and Geophysics.* **35:7-8,** 131-145 (1994).
31. Kotov A.B., Kozakov I.K., Bibikova E.V., et al.. Duration of episodes of the regional metamorphism in areas of polycyclic development of the endogenic processes: results of U-Pb dating, *J. Petrologiya.* **3:6,** 622-631 (in Russian) (1995).

Proc. 30th Int'l Geol. Congr., Vol.7, pp. 83-94
Xu Zhiqin, Ren Yufeng and Qiu Xiaoping (Eds.)

Uplifting and Post-Orogeny of the Ancient Metamorphic Terrain in Southern Liaoning, East China

LI HAIBING XU ZHIQIN ZENG LINGSEN
Institute of Geology, Chinese Academy of Geological Sciences, Beijing, 100037,P.R. China

Abstract

Since Mesozoic, the area of South Liaoning Province, Northeast China, had undergone defferent intensive tectonic movement: 1. Major Indosinia-Early Yuanshania orogeny (220 — 130Ma). The large scale NWW-SEE trending contraction result in the decollement with sense of NWW between the Archean metamorphic basement and the cover (Pt_3-Pz), accompanied by formation of the syn-tectonic granite; 2. Post-orogeny (120 — 100Ma): The crust of southern Liaoning had experienced extension, which is represented by (a) formation of brittle-ductile normal faults at different depth; (b) Development of the gravity-sliding structure; (c) Well developed SN trending fault basement; (d) Emplacement of the NNE trending granite-porphyry within the metamorphic basement; (e) Eart-west trending compression folds and northward thrusting; (f) Uplifting of the ancient metamorphic basement.
The events causing the ancient metamorphic basement to elevate occurred at the stage of post-orogeny. Transfer from contraction to extension, transformation of thermal and stress regime may lead to uplifting of the basement. Shear heating during the compression decollement may soften the crust. The penetration of a large number of felsic veins into the mylonite foliation would cause the expansion of the crust, under the condition of a number of crystal dislocations and the fluid (non-deep source). and resulted in dynamic partial melting. The softening and melting hot crust is easy to uplift, which cause the transformation from contraction to extension. Retrograde metamorphism, formation of quartz veins and the ductile-brittle transition had accompanied the cooling of the hot uplifting crust. During the transformation, σ_1 (the maximum stress) Changed from NWW-SEE trending (main orogeny) to NNE-SSW trending (port-orogeny), and then vertical (at the late port-orogeng).

Key words: Uplifting, Port-orogeny, contraction, extension, felsic veins, dynamic myrmekite

INTRODUCTION

The crust of southern Liaoning (southern East Liaoning) of the Mesozoic intra-continental mountain chain in the eastern part of the North China plate has a very complex feature of deformation[12,13]. Large-scale taphrogeny in the crust of southern Liaoning that consists of Paleozoic cover and older metamorphic basement took place mainly in the Mesozoic, in successive stages of contraction and extension, i.e. the early orogenic stage (main orogenic stage) and post-orogenic stage. The strong structural deformation in this area has attracted much attention. Cui Keying et al. noted in 1986 that there was a decollement between the Archean basement and cover of southern Liaoning in the Mesozoic and at the same time they found the ESE-striking (100-120°) lateral stretching lineation. Xu Zhiqin et al. in 1991 further confirmed the Mesozoic decollement and stressed that the detachment-overthrust regime and the lag normal fault represented respectively compression and extension of two different structural stresses.

The study of the post-orogeny in the intra-continental belt is an important part of the study of orogenic belt and is significant. The uplifting of the ancient metamorphic basement after orogeny is an important mechanism of post-orogeny.

GEOLOGICAL BACKGROUND

The study region is located in the south of the NNE-striking fold mountain system of eastern Liaoning. The strata here consist of the metamorphic basement composed of the Archean (the Anshan Group, 2,590Ma) of amphibolite facies and the Neoproterozoic (Liaodong Group, 1,668-2,156Ma) of amphibolite-green schist facies and Neoproterozoic-Paleozoic cover [1,2,12]. Generally the 8-10km-thick Neoproterozoic (including the Qingbaikou system and the Sinian) and the Cambro-Ordovician are stable sedimentary cover of the North China platform, but the cover in southern Liaoning was strongly deformed and regionally metamorphosed (green schist facies) in sharp contrast with that of western Liaoning (the vicinity of the Fuzhou Bay) and northern Liaoning where the cover is more stable in structure characterized by very gentle folds and brittle faults and is almost unmetamorphosed. Beside the strongly deformed cover, among the most spectacular are the Archean metamorphic basement appearing as dome running NNE and the vast Indosinian and Yanshanian granites (Fig. 1). It can be seen in Fig. 1 that the ancient metamorphic terrain is cut on the west by the Jinzhou Fault and covered on the north by a Cretaceous fault basin.

Since the Mesozoic, southern Liaoning has undergone at least the early orogeny, a phase of contraction and the post-orogeny, the phase of extension.

EARLY OROGENY----EVENT OF CONTRACTION AND DECOLLEMENT

Studies suggest that the amphibolitic mylonite about 1km thick at the top of the Archean metamorphic basement of amphibolite facies resulted from the large-scale compressional detachment event between the Neoproterozoic-Paleozoic cover and the basement in the initial Mesozoic [12]. As a result of later uplifting and erosion, the mylonite zone encircles the basement. Felsic veins are rampant and WNW-ENE striking stretching lineation is developed pervasively on the foliage plane. Determinations on the microstructures such as S-C structure, asymmetric porphroclast , asymmetric shear folds, etc. suggest that there is westward shearing vergence on both the east and west of the dome (Fig. 2a, b).

The upper cover sheet composed of clastic and carbonate rocks shows continuous, progressive deformation and metamorphism and uniform shear vergence, suggesting that the cover and metamorphic basement together experienced a westward deep detachment. The fact that the detachment was coupled with folding and overthrusting affirmed that the crust accreted through shortening and thickening and the decollement took place under the background of crustal shortening.

On the islands of Guanglu, Changshan, Zhangzi and Haiyang in the Yellow Sea on the east of the Liaodong Peninsula, the lenses composed of the Liaohe group (Pt_1) between the mylonite on top of the Archean basement and the Neoproterozoic cover and the strong shearing characterized by the appearing of cyanite, garnet, phyllite, etc. suggest that the Liaohe group was involved in the same structural event (Fig. 2) (Xu Zhiqin et al., 1994).

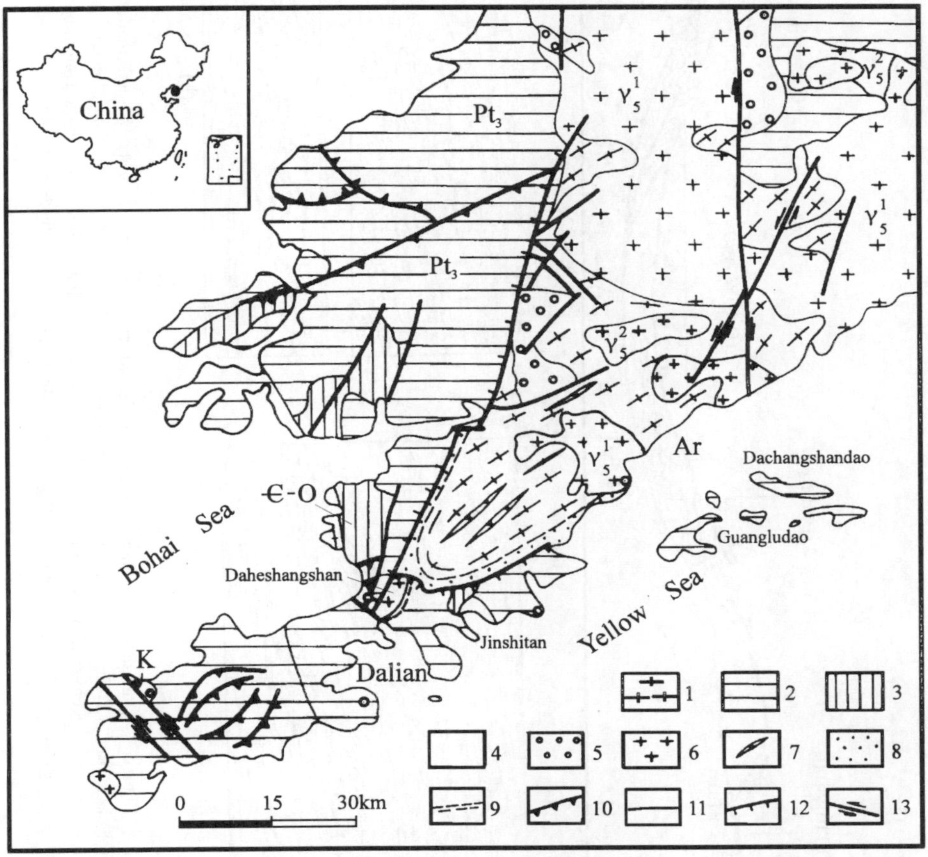

Figure 1. Sketch structural map of southern Liaoning, East China
1. Archean metamorphic basement, 2. Neoproterozoic sedimentary cover, 3. Cambrian-Ordovician sedimentary cover, 4. Carboniferous, 5. Cretaceous sedimentary basin, 6. Granite, 7. Granite-prophyry vein, 8. Mylonite zone, 9. Ductile normal shear zone, 10. Thrustfault, 11. Inferred fault, 12. Normal fault, 13. Stribe-ship fault

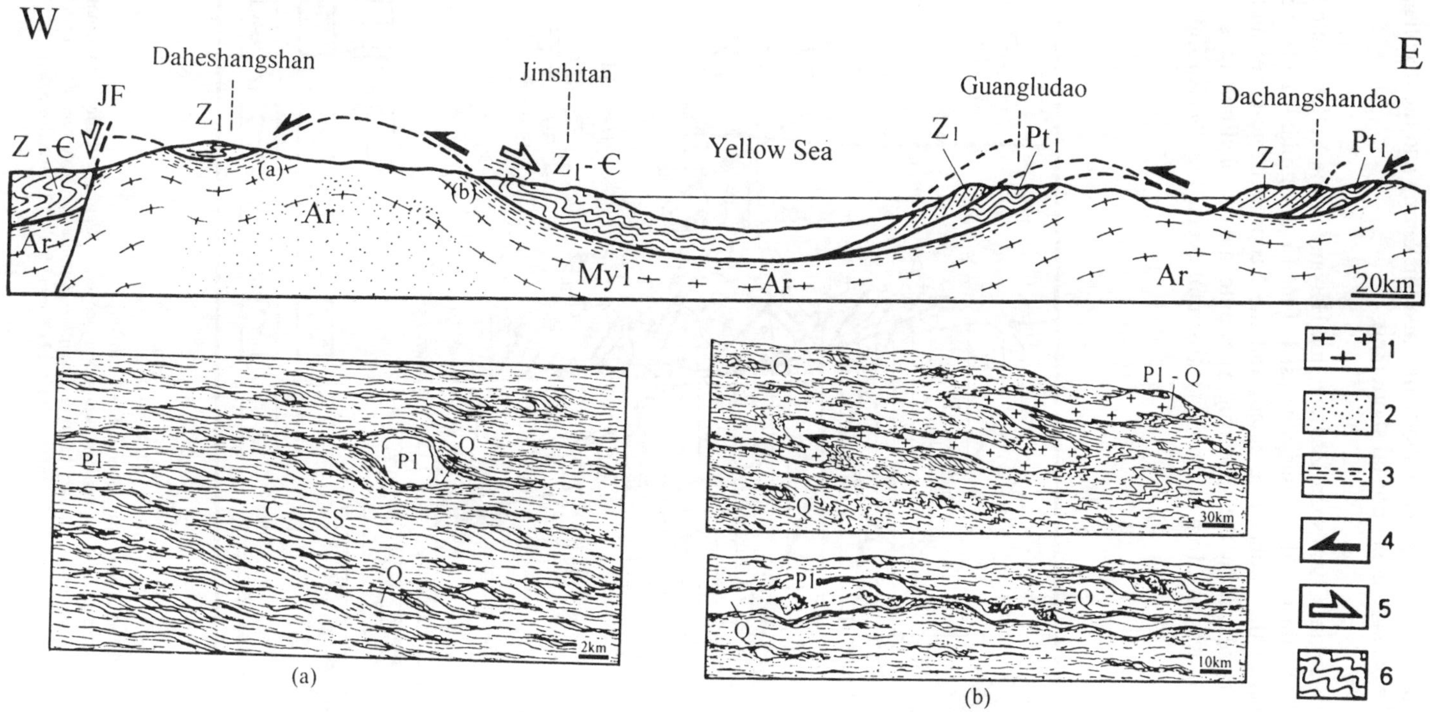

Figure 2. Daheshangshan-Dachangshan dao Island cross section of structural deformation in southern Liaoning. 1. foliation; 2. locale remelting; 3. mylonite (Myl); 4. direction of decollement; 5. direction of extension; 6.cover fold.a-shearing in the Daheshangshan mylonite; b-shearing in the Dongjiagou mylonite;S. C -orientation of S-surface and C-surface; Q-quartz; Pl-plagioclase; Pl-Q-felsic vein

The shortening in the cover was marked by a lot of overthrusts, e.g., the Pulandian-Wafangdian overthrust, in which the Precambrian was thrust on the Early Jurassic coal series [12,15], and the overthrusts in southern Dalian shown in Fig. 1.

Based on the 225Ma age (K-Ar) of the muscovite in the felsic vein in the amphibolitic mylonite [12], the 183-127 Ma age (K-Ar) of the muscovite in the mylonite of the Archean basement [12], 200Ma Indosinian granites occurring in the west and north of the amphibolitic mylonite and the thrusting of Precambrian sequence over the Lower Jurassic, we infer that the age of the decollement-overthrust event is Indosinian-Early Yanshanian.

POST-OROGENY----EXTENSION

The general extension of the cover and basement superimposed on the earlier contractiom decollement and is obviously related to the uplifting of the metamorphic terrain.

In the upper part of the Xiangshuisi mylonite zone in the west of the Daheishangshan Mnt., the Neoproterozoic Qingbaikou system is characterized by the phyllonite and carbonate mylonite of about 50m wide, in which the foliage appears as east-dipping, listric, ductile normal fault, stretching lineation is nearly E-W striking, calcite and quartz porphroclast show undulatory extinction, and calcite, muscovite and sericite make up strip-like S-C structure. Asymmetric swerving of fragmented phenocrysts suggests that the upper layer moved toward the west. Brittle normal faults are widespread in the other members of the cover.

Fig. 3 shows the cross section in the west of the Daheshangshan Mnt. We collected specimens in various parts of the section for sectioning and, by measuring the preferential direction of the C axes of quartz of different occurrences on the Fedorov stage, obtained the following information: ① There are strongly deformed felsic veins in the gently west-dipping mylonites and they were cut by later veins. The quartz fabrics shown in Fig. 3a-d suggests westward slipping as observed in the field; ② In the lower part of the mylonite zone at the center of the dome (Fig. 3e, f), the K feldspar in the veins is coarser, and the veins are thicker and redder and cut across earlier ones. The quartz fabrics in Fig. 3e is a mixture of high-temperature and low-temperature fabrics. The high-temperature fabrics indicates westward slipping while the low-temperature fabrics indicates eastward normal slipping; ③ Fig. 3f-h show the quartz fabrics pattern in the east of the dome in a environment of reddish, strongly deformed felsic veins and granite, in which local dynamic remelting was extensive and dynamic mymekite structure is obvious. Fig. 3f shows the low-temperature fabrics and that the westward movement was more important. The eastward movement indicates that contraction started to turn to extension. Fig. 3g suggests the medium-temperature fabrics of mainly eastward movement. The specimen of Fig. 3h was collected in the deformed granite and the diagram shows the pattern of high-temperature fabrics and movement of two directions; ④ The fabrics of the fragmented quartz phenocrysts (Fig. 3i, j) in the carbonate mylonite at the bottom of the cover sheet demonstrates the normal eastward slipping under medium-low temperature.

The above analysis demonstrates that the felsic veins in the mylonite were formed under medium temperature, and the felsic veins and deformed granite at the center of the dome were formed under higher temperature and both suffered low-temperature deformation. Analysis on the quartz fabrics suggests that there was an earlier westward detachment followed later by a bilateral normal slipping on both sides of the dome (westward slipping on the west and eastward on the east).

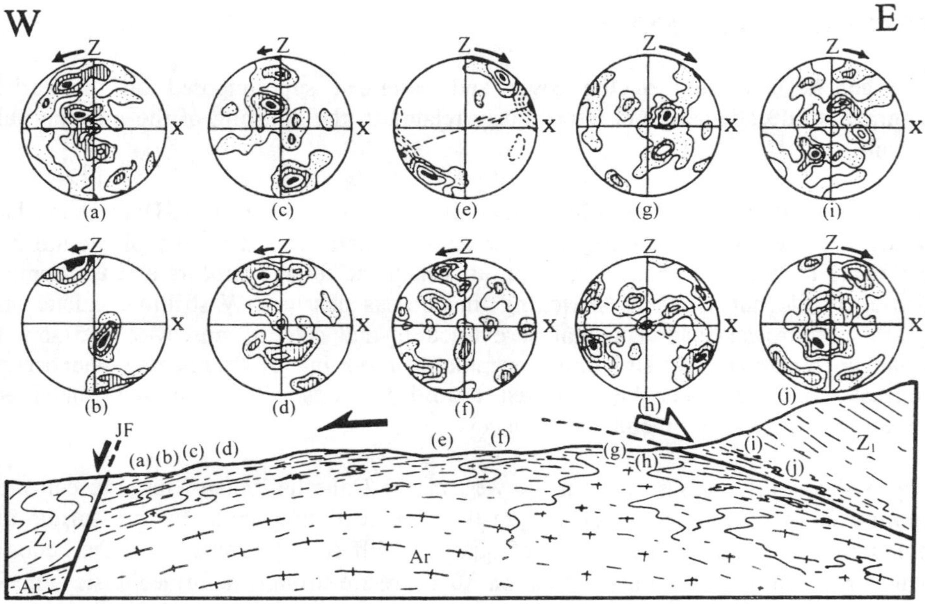

Figure 3. Structural section and quartz fabrics in the west of the Dashangshan in southern Liaoning Ar-metamorphic rocks of the Archean Anshan group; Pl-Q-felsic veins; Myl-mylonite; Z_1-Lower Sinian; Z_1-Є -Lower Sinian-Cambrian; γ_5^1 Indosinian granite; solid arrows show directions of decollement and hollow arrows the direction of normal slipping of ductile and brittle normal faults. (a), (b), (c), (d), (e), (f), (g), (h), (i) and (j), sites of oriented specimens collected and corresponding diagrams of preferential directions of quartz. X-direction of maximum extension; Z-direction of maximum compression; the small black arrows in the diagrams of preferential directions indicate the directions of shearing. (a) structurally recrystallized quartz (*d*=0.3mm ±), 102 grains, 1-2-4-7%; (b) structurally recrystallized quartz (*d*=0.2-0.4mm ±), 100 grains, 1-2-4-7%; (c) mylonitic quartz (*d*=0.05-0.1mm ±),102 grains, 1-2-4-7%; (d) structurally recrystallized quartz in the quartz veins (*d*=0.3-0.4mm ±),132 grains, 1-3-7-9%; (e) quartz phenocrysts (*d*=0.4-0.5mm ±), 70 grains, 1-3-5-6%; (f) vermicular quartz (*d*=0.2-0.4mm ±),84 grains, 1-3-5-7%; (g) mosaic quartz (*d*=0.44mm ±) 105 grains, 1-3-5-9%; (h) structurally recrystallized quartz in granite (*d*=0.3-0.4mm ±), 150 grains, 1-3-7-9%; (i) fragmented quartz phenocrysts in carbonate mylonite (*d*=0.3-0.4mm ±), 100 grains, 1-2-3-5%; (j) fragmented quartz phenocrysts in carbonate mylonite (*d*=0.3-0.4mm ±), 100 grains, 1-2-3-5%.

The structural section of the crust of southern Liaoning (Fig. 2) shows the occurrence of the Archean metamorphic dome on the peninsula and under the sea. Beside the Jinzhou dome on the peninsula, there are two smaller domes under the sea, all striking NNE. The earlier formed mylonitic foliage is wavy. According to the presence of normal slipping structure, the low-angle normal shear zone on the east of the dome, and the gravitational slipping structure in Jinshita, the absence of the bottom horizons of the cover, the development of the NNE striking Lower Cretaceous continental fault basin in the region, and the fact that the NNE Jinzhou normal fault made the east side of the metamorphic terrain rise and the west fall, and considering the 114Ma age of the later metamorphic event at the center of the dome (Bi, Ar^{40}/Ar^{39})[12], and the 103-118Ma metamorphic ages of the muscovite in the normal ductile shear zone (K-Ar), we can conclude that, following the Indosinian-Early Yanshanian decollement event (contraction), the crust experienced an extensional deformation after the Early Cretaceous, leading to the formation of the metamorphic dome.

UPLIFTING OF THE ANCIENT METAMORPHIC TERRAIN

The uplifting of the Archean metamorphic basement is closely related to the post-orogeny, the extensional deformation.

NNE striking Cretaceous tensile granite-porphyry veins are widespread in the ancient metamorphic basement and ptygmatic folds of felsic veins of plastic flow, tiny normal

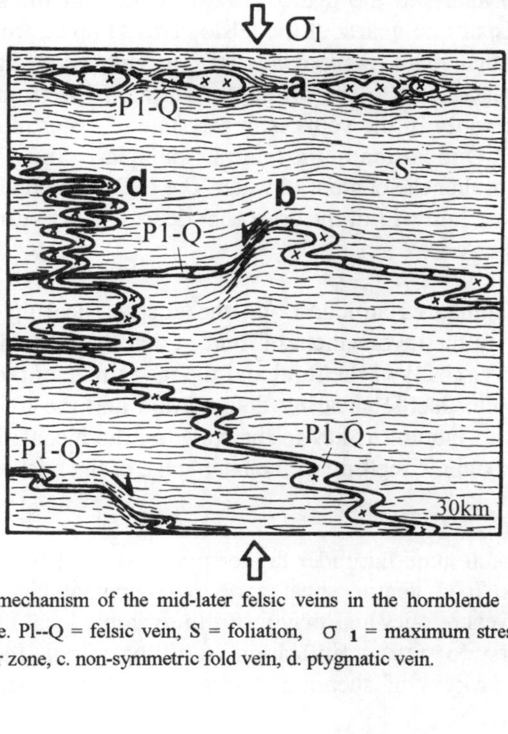

Figure 4. Deformation mechanism of the mid-later felsic veins in the hornblende gneiss at the core of the Jinzhou metamorphic dome. Pl--Q = felsic vein, S = foliation, σ_1 = maximum stress, a. extended boudinage vein, b.ductile normal shear zone, c. non-symmetric fold vein, d. ptygmatic vein.

ductile shear zones, and boudinage structures of felsic veins occur near the center of the basement, suggesting that they were formed when the maximum stress (σ_1) is vertical(Fig.4). The uplifting of the basement is intimately associated with the felsic veins and granite.

Function of fluid in the shearing deformation

Syn-tectonic felsic, pegmatite and migmatite veins are enormous in the mylonite zones in the Jinzhou uplift of southern Liaoning, indicating that dynamic local melting took place in the course of shearing deformation, forming a crystal-fluid two-facies system. The size of the quartz grains in the felsic veins are generally smaller than the size of the dynamically recrystallized quartz in the host rocks, and this is probably that the deformation of quartz is related to the melt produced by dynamic local melting. The melt intensifies the deformation of quartz, prompting the transformation of the deformation mechanism, displacement creeping, of quartz to the melt-intensified grain boundary slipping in accordance with diffusing,while the numerous brittle cracks in the feldspar are related to the thermal expanding effect of filling melt. The deformation features of the felsic veins demonstrate that, with the involvement of fluid, deformation is very sensitive to the grain size, and dynamic recrystallization depends more on the diffusing transfer of matter than on the plasticity within the crystal. The formation of the metamorphic differentiation bedding in the amphibolitic mylonite marked by the rhythm of the syn-detachment shearing veins is also closely related to the fluid.

Using FT infrared spectrophotometer under the scanning condition of 400-4000cm^{-1}, Xu Zhiqin et al. (1984) analyzed the hydroxyl radical (OH) of the sericite and the fluid inclusion in the feldspar and quartz of the felsic veins sampled from the study area and demonstrated that the main components of the inclusion are water and some carbondioxide, and according to the infrared absorption spectrum and the position of absorption peak, the water in the veins is in the state of water molecule. However, the relative contents of H_2O inclusion in the feldspar and quartz and (OH) in the sericite are different in that the higher the content of water, the higher the relative content of (OH). This indicates that the fluid within the crust (not deeper source) directly took part in the formation of the felsic veins.

The deformational laminated structure and the ductile shearing in the lower-middle crust is associated with the involvement of fluid in that the formation of the structure is closely related to the viscosity of the fluid and the magnitude of the shear stress[6]. The function of fluid in the ductile shearing deformation and the following post-orogenic extension in southern Liaoning is key to the study on the mechanism of the retrograde metamorphism and dynamic melting in the later stage of the ductile shearing.

Dynamic local melting

Local melting of crustal material under higher rate of strain has become a consensus of structural geologists. In a ductile shear zone, a competent layer may produce more shearing heat than weaker ones and may result in dynamic local melting. Metamorphic differentiation refers to the stratification resulting from the recombination of components in the process of shearing deformation. It was believed that the well-developed felsic veins and their deformation in the ductile shear zones in the crust of southern Liaoning resulted from the local melting in the shearing deformation. Ji Shaocheng (1988) pointed out that numerous felsic veins in the ductile shear zones are indicator of shearing heating and dynamic local melting in the ductile shearing deformation. In the amphibolitic mylonite in southern Liaoning, the widespread metamorphic differentiation bedding of alternating black and white strips, the white consisting of felsic veins and the black of oriented melanocratic minerals, is a joint result of mechanical force, chemical reaction and resolving transfer [3], and is a self-organization phenomenon. During the shearing deformation, the shearing stress and local temperature held a dynamic, non-linear relationship, i.e., the shearing stress is dynamically changing with respect to time [16].

Myrmekite and rudimentary vermicular quartz are well-developed in the felsic veins in this region and are obviously controlled by stress and strain, and are the product of deformational reaction, so this structure with vermicular quartz is called dynamic myrmekite structure(Fig.5) [13]. myrmekite is a mutual growth of plagioclase and vermicular quartz, small-grained and with graphic texture, indicating that its components are eutectic and are the crystallization of the inter-crystal melt (crystal mush) produced in partial melting, and its shape and small-grained texture suggest that this flow structure under strong plastic deformation is very probably relevant to the function of fluid.

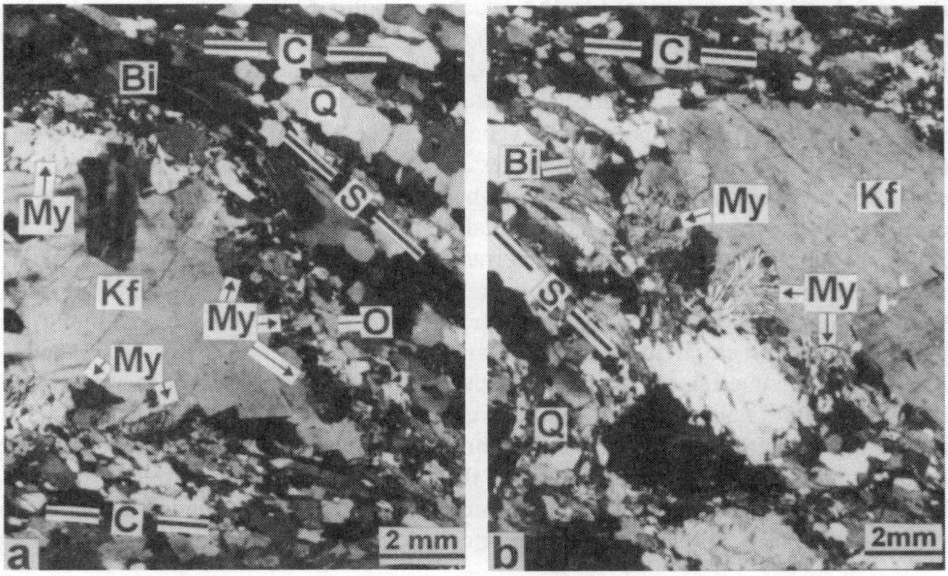

Figure 5. Optical photomicrographs of samples LS21 and 9118 (Southern Liaoning).Bi=biotite; kf=k-feldspar; O=oligoclase; Q=quartz; My=myrmekite. k-feldspar megacryst between S-and C-surfaces. The myrmekite (My) formed on margins of k-feldspar along S-C-surfaces (a), and S-surface (b) . The myrmekite are obviously controlled by stress and strain, and are the product of deformational myrmekite structure.

The dynamic myrmekite structyre occurred in the felsic veins of the detachment mylonite zone while the veins grew in the whole history of compressional detachment and extensional deformation, thus, the study of the background, thermodynamic conditions and mechanism of their formation is very important.

The appearance of myrmekite in the deformed rocks of shear zone can be taken as the result of the metasomatism in the later stage of local melting brought about by shearing, and right at this time, compressional deformation had turned to extensional. The temperature of metasomatism should be lower than that of remelting, otherwise, oligoclase and quartz will be eutectic at higher temperature. According to calculation, the temperature of metasomatism in the rocks of a shear zone is about 440-540 ℃ [7-11].. The texture of the vermicular quartz in the mylonite of southern Liaoning suggests that the temperature of their formation is about 400-450 ℃, while that of the earlier myrmekite may be higher than 400-450 ℃.

Turn from compression to extension

The felsic veins are the result of the dynamic local melting with the involvement of fluid, while the dynamic mymekite structure is the result of the metasomatism of the dynamic local melting and can be taken as the mark of the transfer from contraction to extension [13].

The mechanism triggering the uplifting of the ancient metamorphic terrain is the turn caused by the shearing heating effect from the compressional deformation to extension. Because shearing heat produced by the detachment softened the crust and owing to the involvement of fluid (not from deep source) and the enormous displacements within the lattice of the minerals of the deformed rocks, local melting (i.e. dynamic local remelting, c. 450-550 ℃) in the crust might occur at the temperature lower that the normal melting point, and a lot of felsic and granitic veins filled the ever existing crevice along the foliage of structurally incompetent zones (mylonite zones), leading to the volume expansion and thermal expansion of the whole crust. Thus, the softened, remolten hot crust would easily rise if triggered by a certain force (compressing ?) from the depth and right at this time compressional deformation turned to extension, and dynamic myrmekite and vermicular quartz were formed because of the metasomatic reaction between the crystals in the crystal mush produced by dynamic remelting or along the existing micro-structural planes, and meanwhile, the temperature was lower down, leading to the superimposing of low-temperature fabrics of quartz on medium- and high-temperature fabrics, the formation of quartz veins, retrograde metamorphism and the gradual transformation from ductile to brittle deformation. Thus, we can believe that this "dynamic remelting" thermal dome of the ancient metamorphic terrain in southern Liaoning is different from the "granite" thermal dome (e.g., the Yajiang thermal dome in western Sichuan, [13]) and metamorphic core complex.

In the structural evolution in southern Liaoning, the time interval of the turn form compressional deformation to extension is about 10-20Ma. Analysis on the paragenous mineral assemblages of the mylonites shows that the Archean metamorphic basement experienced an amphibole-biotite-chlorite, epidote retrograde metamorphism. Beside the fluid in the shearing deformation, the retrograde metamorphism from amphibolite facies to low green schist facies is closely related to the basement uplifting. Calculation of temperature and pressure of metamorphic facies gives a lower rate of basement uplifting. Thermodynamic calculation of Glazner et al. (1985) on overthrust belt shows that the sharp strengthening (once the overthrusting thickening takes place) of the lithosphere will turn to sharp weakening (10-30Ma after overthrusting), while Sonder (1987) believes that the time interval of the turn from the termination of compression in an orogenic belt to the start of extension is relevant to the initial temperature of the Moho, in that the higher the temperature, the shorter the interval. The large-scale detachment-overthrusting in the crust of southern Liaoning prompted the crustal shortening and vertical thickening, resulting in gravitational instability. The strong granitic magmatism in the process of ductile shearing suggests that the initial Moho temperature in southern Liaoning was relatively high, and thus, the basement uplifting would take place in a shorter time interval, leading to crustal extension. The time interval of the turn in southern Liaoning depended on the thickness resulting from contraction and the syn-tectonic magmatism.

CONCLUSION

Studies on the deformation in southern Liaoning demonstrate that (1) During the early orogeny (220-130Ma) large-scale WNW-ESE contraction caused WNW decollement between the metamorphic basement and cover marked by the amphibolitic mylonite zones. At the same time there was intra-continental syn-tectonic granite; (2) During the post-orogeny (120-100Ma), crustal extension took place, leading to the formation of ductile, ductile-brittle and brittle normal faults of different orders, and the formation of fault basins. The formation of gravitational sliding structure resulted in the uplifting of the ancient metamorphic basement with the intrusion of NNE striking granite-porphyry veins; (3) Dynamic local melting in the process of shearing deformation was the result of shearing heating, the felsic veins were produced by dynamic local melting with the involvement of fluid and the enormous felsic veins in the mylonite zones were produced in the process of decollement and further experienced extensional deformation; and (4) The dynamic myrmekite structure was the result of metasomatism of dynamic local melting and can be taken as the indicator of the transfer from contraction deformation to extension.

Acknowledgments:

We are grateful to Senior Engineer Li Dianzhi, Engineer-in-chief Meng Qingcheng and Senior Engineer Zhang Guoren of the Regional Geological Survey of Liaoning for their great help during our field work.

REFERENCES

1. Bureau of Geology and Mineral Resources of Liaoning Province. *Regional Geology of Liaoning Province*. Geological Publishing House, Beijing, 380 (1989).
2. K.Y.Cui, Q.B.Pang and Y.K.Xu. On principal feature of ductile nappe structure in Southe Liaoning. *Bulletin of the Geological Society of Liaoning Province*, **2**, 45-66(1986).
3. T.Dewers and P.Ortoleva. Differentiated structures arising from mechano-chemical feedback in stressed rock. *Earth-Sciences Reviews*, **29**, 283-298(1991).
4. A.F.Glazner and J.M.Barley. Evolution of lithospheric strength after thrusting. *Geology*, **13**, 42-45(1985).
5. S.C. Ji. Tectonical implications of partial melting (I): an experimental study of deformation mechanism transition. *Scientia Geological Sinica*,**4**,347-356(1988).
6. R.Meissner. Rupture, creep, lamellae and crocodiles: happenings in the continental crust. *Terranova*, **1**, 17-28 (1989).
7. K.O'Hara and L.P.Gromet. Two distinct Precambrian (Avalonian) terranes in southeastern New England and their late Paleozoic juxtaposition. *American Journal of Science*, **285**, 673-709 (1985).
8. C.Simpson. Evidence for deformation-induced k-feldspar replacement by myrmekite. *J.metamorphic Geol.*, ,261-275 (1989).
9. M.Sonder, P.C.England and B.P.Wernicke. A physical model for Cenozoic extension of western North American. *Geological Society Special Publication*. London, **28**, 187-201 (1987).
10. J.C.Stormer. A practical two feldspar geothermometer. *American Mineralogist*, **60**, 667-674 (1975).
11. J.A.Whitney and J.C.Stormer. The distribution of $NaAlSi_3$ between coexisting microcline and plagioclase and its effect on the geothermometric calculations. *American Mineralogist*, **62**, 689-891 (1977).
12. .Z.Q.Xu, H.B.Li, Z.X.Wang and D.Z.Li. Crustal contraction and extension in Southern Liaoning. *Geological review*. **37**, 193-202(1991).
13. .Z.Q.Xu, H.B.Li and L.H.Guo. Dynamic Myrmekite, Dynamic Melting and tranfer from Crustal contraction to extension: Discussion on Uplifting Mechanism of Ancient Metamorphic Terrane in Southern Liaoning Province, In: *Extension*; edited by Qian XiangLing, Geological Publishing House, Beijing, China. 109-119(1994).
14. .Z.Q.Xu, L.W.Hou, Z.X.Wang, X.F.Fu and D.K.Wang. *Orogenic processes of the Songpan-Garze orogenic belt of China* Geological Publishing House, Beijing. 1-190(1992).
15. L.S.Zeng and H.B.Li. Structural Deformation and Fluid Process in Jinzhou Uplifting Region, South Liaoning Praince. *GeologicalJournal of China Universities*. **2**, 155-165 (1996).
16. Y.Q.zhu and Y.L.Shi. Shear heating and partial melting of granite-Thermal structure at overthrusted terrains in the greater Himalaya. *Acta Geophysica Sinica*, **33**, 408-415 (1990).

Proc. 30th Int'l Geol. Congr., Vol.7, pp. 95-120
Xu Zhiqin, Ren Yufeng and Qiu Xiaoping (Eds.)

Basic Regularities and Driving Mechanisms of the Neotectonic Crustal Uplift

EUGENE V. ARTYUSHKOV[1,2] and ALBRECHT W. HOFMANN[3]
[1] *United Institute of Physics of the Earth, B. Gruzinskaya, 10, Moscow, 123810, Russia*
[2] *Paleogeophysics & Geodynamics, Kräftriket 24, Stockholm University, 10691, Sweden*
[3] *Max-Planck-Institut für Chemie, Mainz, 55020, Germany*

Abstract

According to a large volume of data, a rapid crustal uplift began in the Oligocene over about one half of the continental areas after a period of relative stability, ~150 Ma long. This Neotectonic uplift has produced most of the present high topographic features, e.g., the South African plateau and many mountain ranges in the Alpine belt and Central and North-eastern Asia. A wide occurrence of shallow-water fossils indicates, that the Tien Shan and Caucasus were below sea level 40 Ma and 10 Ma ago, respectively. A strong acceleration of the uplift took place during the last 3-5 Ma. 1.6 Ma ago the Tien Shan was at half the present-day elevation. Most of the Neotectonic uplift took place on Precambrian, Paleozoic and Mesozoic crust and occurred far from active plate boundaries, without significant horizontal deformation and disturbance of isostasy. High gradients of the uplift are observed in many areas. Under such circumstances, the uplift should be caused by a density decrease in the lithosphere. Geophysical data show that in some regions the Neotectonic uplift was associated with recent asthenospheric upwelling which resulted from a partial or complete erosion of mantle lithosphere. This collapse of mantle lithosphere occurred without significant horizontal deformations. It could probably be associated with infiltration of volatiles into this layer from a fluid-containing anomalous mantle which ascended from a deep mantle in some places and rapidly spread into the other areas along the base of the lithosphere. Asthenospheric upwelling can ensure up to 1.5-2.5 km of the uplift. Delamination of dense eclogite from the crust or transformation of garnet granulites in the lower crust into a less dense gabbro under a contact with the asthenosphere can increase the magnitude of the uplift up to 3-4 km.

Keywords: Neotectonic uplift, southern Africa, Tien Shan, Verkhoyansk range, Alpine fold belt, asthenospheric upwelling

INTRODUCTION

An elevated topography is typical of a considerable part of continental areas. The highest continent is Asia; 28% of it is above 1 km and 13% higher than 3 km with the altitudes of ~4-5 km on the Tibetan Plateau, in the Himalayas, Karakoram, Hindu Kush, Kunlun Shan, Pamirs and Tien Shan [15]. Denudation of a high topography occurs at a high rate in humid areas [94]. Even taking into account the isostatic uplift in response to the denudation, high mountains should be destroyed in such areas in a few tens of million years or more rapidly. This indicates that most of the present mountains have formed recently. The same conclusion follows from a large volume of geomorphological, geological, paleobotanical, paleontological, paleosoil and fission track data. They show that a rapid crustal uplift began in many regions since the Oligocene after a period of relative stability, 100-150 Ma long [9, 17, 44, 51, 54-58, 66, 68-71, 73, 74, 83, 91, 94, 101-103]. This took place, on, e.g., the Tibetan, South and East African plateaus, in mountain ranges of central and north-eastern Asia and in the Alpine belt, in the Andes and North American Cordilleras. In many regions the uplift was associated with magmatism (e.g., on the Tibetan Plateau, in central Asia and in many regions in Africa). The crustal uplift which took place during the last several tens of million years is usually called "the Neotectonic uplift". In different areas its onset ranges from the Late Eocene to the Pleistocene. A strong acceleration of the Neotectonic uplift took place during the last 5 Ma when most of the present high topographic features have been formed.

The Neotectonic crustal uplift is considered as a cause of dramatic changes in the climate, oceanic and atmospheric circulation, and glaciations which have taken place during the last several million years [79, 82]. According to an opposite point of view, the uplift took place due a rapid denudation of the pre-existing topography which was enhanced by the climatic changes [62]. It has been, however, shown that this effect cannot be responsible of the formation of some present high plateaus and mountain ranges [32]. Most scientists working in a field of geodynamics are unaware of the intensity of the Neotectonic crustal uplift on a global scale and its role in formation of the present high topography on the continents. Controversial ideas exist about physical mechanisms of the uplift (see below). In this paper we present several examples of the Neotectonic uplift and discuss its characteristic features and possible driving mechanisms.

TYPICAL EXAMPLES

Southern Africa

This area (Fig. 1) is underlain by crystalline crust of predominantly Archean and Middle Proterozoic age. Its geomorphic evolution in the Mesozoic-Pleistocene has been studied in detail [39, 74]. In the Middle Jurassic southern Africa was located at a considerable altitude in the inner part of Gondwana. The crustal uplift had taken place on its western, southern and eastern margins before the continental break up which occurred in the Late Jurassic. After the break up these regions stood high above the adjacent ocean. Their denudation by backwearing produced a steep scarp — the Great Escarpment, 300-1000 m high, which is located 50-250 km from the coast [73].

In Fig. 2 this scarp, ~1 km high, can be seen at points C and D. During the Cretaceous and Paleogene a planation surface the "African surface" formed below the escarpment (see Fig. 1). It was slowly rising from the ocean toward the escarpment. Planation surfaces which have the ocean as a basis of erosion and form near to it are located at a very low altitude. The African surface below the escarpment could not be higher than 100-200 m near the scarp. Above the Great Escarpment — on the wide interior plateau, another sub horizontal planation surface of the African age formed in the Cretaceous and Paleogene. It was several hundred meters above the synchronous — African planation surface which was located at a low altitude below the escarpment — on the margins of southern Africa. The African planation surface is covered by a weathered mantle, up to

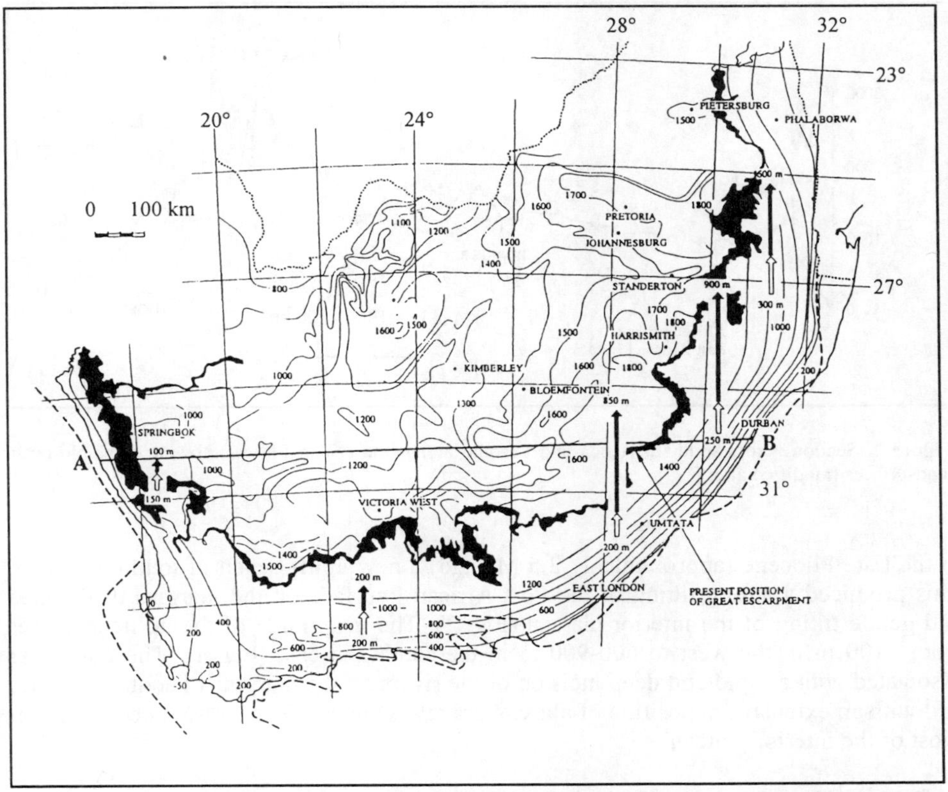

Figure 1. The altitude of the African planation surface (in meters, modified after [74]). Open arrows show the magnitude of the Miocene crustal uplift in meters. Solid arrows give the uplift in the Late Pliocene-Pleistocene. AB — the line of profile of Fig. 2.

60-70 m thick, which formed during a period of time, 100 Ma long. The formation of this weathered layer required a slow removal of ~1 km of rocks. This shows that a slow crustal uplift took place in southern Africa in the Cretaceous – Eocene. The products of erosion were deposited on the adjacent shelf and continental slope. By Oligocene time the formation of the planation surface was completed which is evidenced by a strong decrease in the rate of deposition on the surrounding shelf.

After a period of tectonic quiescence, ~100 Ma long, a moderate crustal uplift began at the end of the Early Miocene ~18 Ma ago. This resulted in a new cycle of denudation, 15 Ma long. By the Late Pliocene a new and rather irregular Post African I surface with an imperfect planation was formed (see Fig. 2). The difference between the African and Post-African I planation surfaces approximately corresponds to the magnitude of the crustal uplift in the Late Miocene. It rapidly increased inland — from the coast toward the axis of the belt where a strong uplift took place before the break up of Gondwana. In the inner part of the subcontinent the magnitude of the uplift slowly increased from 100-150 m in the west to 200-300 m in the east (see also Fig 1).

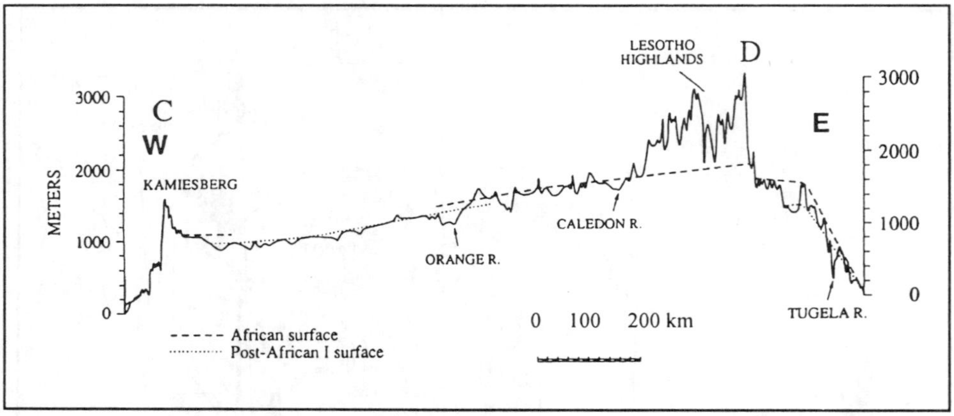

Figure 2. Section across southern Africa and Lesotho along line AB in Fig. 1, showing principle cyclic landsurfaces (modified after [74]).

In the Late Pliocene (approximately 2.5 Ma ago) a new uplift began in southern Africa. This produced a strong tilting of the Post-African I surface on the margins of the area and gentle tilting of the interior plateau (Fig. 2). The magnitude of the uplift increases from ~100 m in the west to 600-900 m in the south and east (Fig. 1). The uplift was associated with a rapid and deep incision of the rivers on the margins of southern Africa and with an extensive deposition of alluvial gravels. Only a minor incision occurred over most of the interior plateau.

Before the Neotectonic uplift the altitude of the African planation surface below the Great Escarpment did not exceed ζ_0 ~100 m in the west and ζ_0 ~200 m in the south and east. This surface is now ζ_1 ~400 m above sea-level in the west and ζ_1 ~1000-1400 m above this level in the south and east (Fig. 1). A drop of sea-level Δh_{sl} ~100 m took place during the last 18 Ma [37]. The total magnitude of the Neotectonic uplift is:

$$\Delta\zeta = \zeta_1 - \zeta_0 - \Delta h_{sl}. \quad (1)$$

It is $\Delta\zeta$ ~200 m in the west and $\Delta\zeta$ ~700-1100 m in the south and east. Southern Africa is surrounded by a shallow water shelf in the west, south and east. Only a small subsidence of ~100-200 m took place there during the Neogene [24]. Hence the area of the Neotectonic uplift was separated by a hinge line from the shelf. No significant horizontal deformations took place in a solid Precambrian lithosphere of southern Africa during the Neotectonic uplift. No intense magmatism occurred in the area after the kimberlite intrusions which took place ~90 Ma ago. The active volcano, however, exists in Namibia [C.D. Ollier, personal comm.] and volcanic eruption took recently place in the east of the uplifted area [T.C. Partridge, personal comm.].

The Tien Shan

The Northern Tien Shan is underlain by the crust which formed after a strong convergence in the Silurian [42]. In the Middle and Southern Tien Shan convergence terminated by the Triassic. After planation of the Middle and Late Paleozoic mountains,

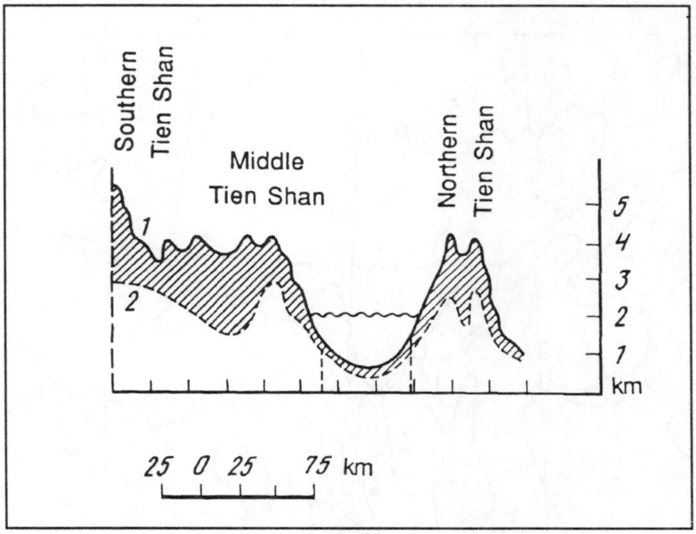

Figure 3. The relief of the Tien Shan, present (1) and about 1 Ma ago (2). Modified after [50].

the area was flooded by a shallow water sea since the Jurassic and until the Eocene [55, 85, 89]. The crustal uplift in the Late Oligocene formed a topography, up to 0.5-1 km high. The uplift was associated with basaltic volcanism.

A slow denudation of low mountains occurred in the Miocene. About 5 Ma ago at the transition from the Miocene to the Pliocene a rapid crustal uplift began. It formed the present Tien Shan which is 4-5 km high on average. The uplift evolved with a strong acceleration in time and has produced up to 30-50% of the present topography during the last ~1 Ma (Fig. 3). The GPS data indicate a present rate of compression of the Tien Shan of 1 cm/a. Unfortunately, they can hardly give a reliable value of the rate of compression, since a large Susamyr earthquake with lateral displacements of ~60 cm took place during the period of observations. A detailed mapping of the deformations of the Paleogene planation surface by bending and thrusting shows that the uplift in the Pliocene and Pleistocene was associated with shortening of the crust by 5% on average and up to 7% in some places [55, 56].

The Verkhoyansk and Chersky ranges

These two mountain ranges are located in the north-east of Eurasia. Strong convergence with intensive shortening of the crust took place between the Jurassic and Cretaceous in the Chersky range and in the middle of the Early Cretaceous in the Verkhoyansk range [5]. No high mountains formed at that time, and the crustal surface remained at a low altitude above sea-level [53]. No more significant convergence occurred in the area. 20 Ma after the end of the Early Cretaceous convergence, a moderate crustal uplift took place at the beginning of the Late Cretaceous. Since the second half of the Late Cretaceous and until the Early Oligocene a lowland with a planation surface existed in the area.

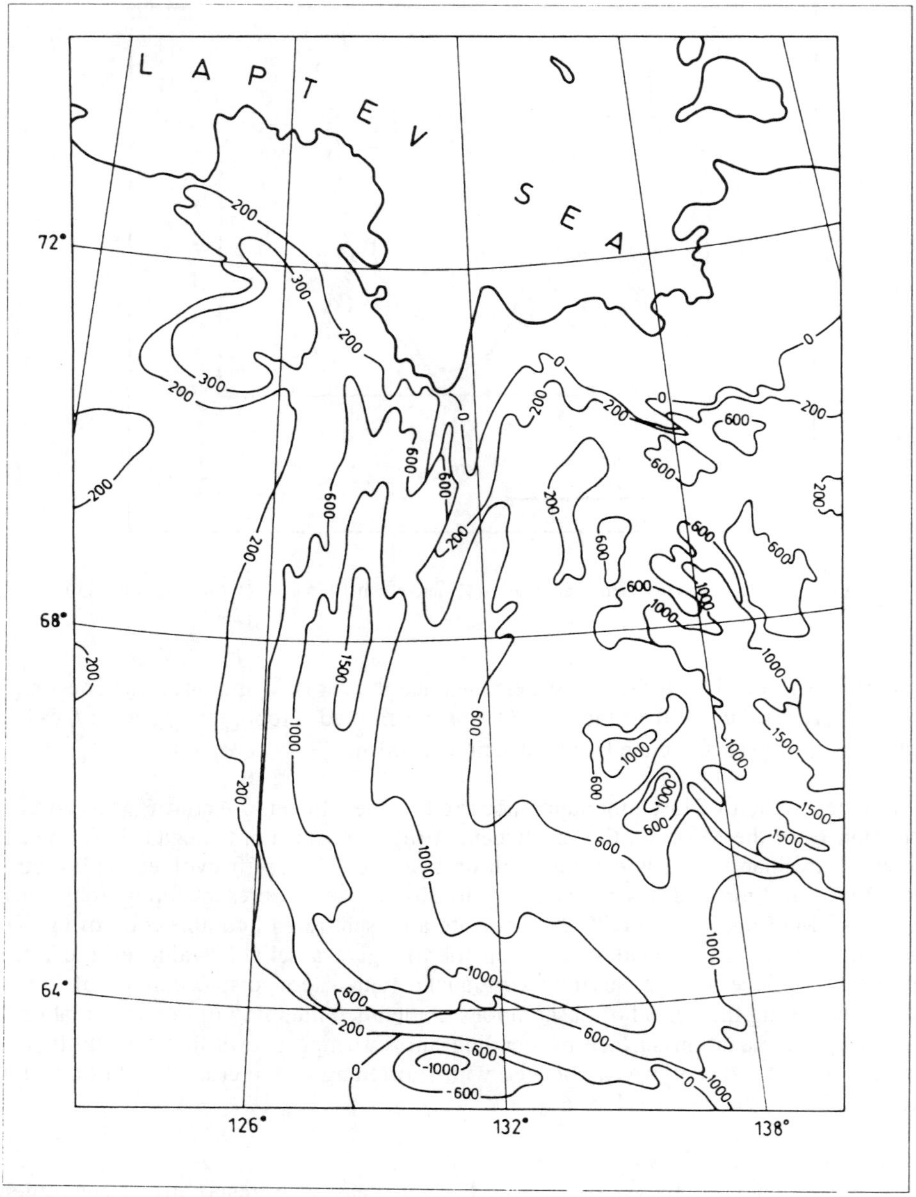

Figure 4. The magnitude of the crustal uplift in the Verkhoyansk and Chersky ranges in the Pliocene-Pleistocene (in meters). Modified after [71]. The Verkhoyansk range is located in the middle part of the Figure between 126° and 132° E. The western part of the Chersky range falls into the right-hand side of the Figure.

A new dome-like crustal uplift occurred in the Late Oligocene. This produced two ranges, up to 1-1.5 km high in the areas of the present Verkhoyansk and Chersky mountain ranges. These low mountains were eroded by the Late Miocene with a formation of a new planation surface at the altitude about 0.5 km. In the Late Pliocene, 3 Ma ago, a rapid crustal uplift began. Its magnitude is shown in Fig. 4. As the preceding uplift in the Late Oligocene, it was more intensive in regions where strong convergence took place in the Mesozoic. During the last 3 Ma the uplift formed the present Verkhoyansk and Chersky ranges. In these regions the Early Pliocene planation surface has been uplifted by up to 1-1.5 km (Fig. 4) and reached the altitude of 2 km. Some mountains reached the height of 3 km. No shortening of the crust took place in the Mesozoic in sedimentary basins which are located between the above mountain ranges. In this area the Early Pliocene planation surface has been uplifted to the altitude of only 1 km. The recent crustal uplift was associated with horizontal deformations of the lithosphere of a few per cent. A slight sub latitudinal shortening took place in the Verkhoyansk range, while slight extension in the same direction occurred in the Chersky range.

Some mountain ranges in the Alpine fold belt

Most of the mountain ranges in the Alpine fold belt were formed by a rapid crustal uplift in the Pliocene and Pleistocene. This is inferred from an increase in the height of numerous river terraces, denudational surfaces and marine deposits with their age. Several typical examples are presented in Fig. 5 which show that the crustal uplift of 1-4 km took place during the last 5 Ma.

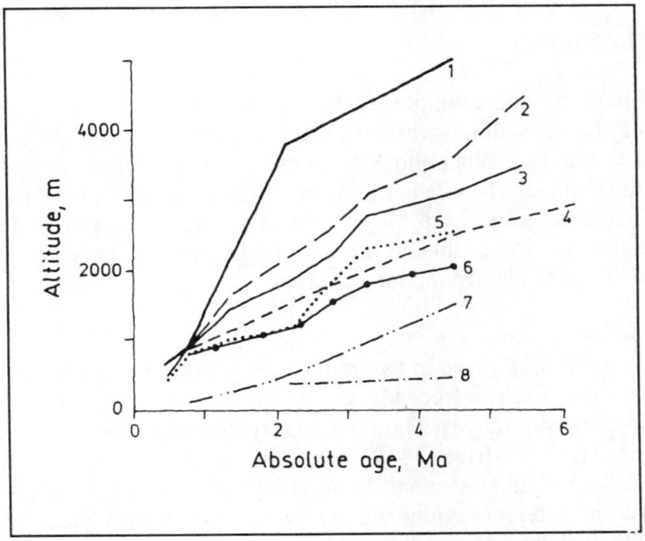

Figure 5. The altitude of denudational surfaces as a function of their age for some mountain ranges in th Alpine fold belt (modified after [54]). 1 — the Central Greater Caucasus, 2 — south-eastern Greate Caucasus, 3 — Lesser Caucasus, 4 — Kopet Dag, 5 — Talysh block in the south-east of the Lesse Caucasus, 6 — Carpathians, 7 — Crimea, 8 — Carpathian foreland.

10 Ma ago the Greater Caucasus was covered by a shallow water shelf. The crustal uplift began in the Late Miocene [57] and formed the mountain range which is up to 4-5 km high. Pre-Middle Sarmatian shallow water deposits are now found on the top of mountains of this height. The uplift is of a dome-like type, and the present ragged topography has been formed due to incision of the rivers. The crustal uplift developed with a strong acceleration, and most of the present topography has formed during the last 5 Ma (line 1 in Fig. 5). A major part of shortening of the crust in the Greater Caucasus had been completed before the onset of the Neotectonic uplift; however, until the Late Miocene the crustal surface remained below sea-level or at a very low altitude above it. Several short episodes of convergence occurred in the Pliocene and Pleistocene [84]. They were associated with no strong additional crustal uplift. Thus, the Neotectonic uplift in the Greater Caucasus was not directly associated with convergence.

A similar situation is typical of the Lesser Caucasus and Kopet Dag. The uplift of the Lesser Caucasus that shown in Fig. 5 (line 3) took place after a predominant part of convergence had been completed in this fold belt [84]. The last convergence which took place 3.5-4 Ma ago was not associated with a significant additional uplift. The major convergence in the Kopet Dag which occurred about 3.5 Ma ago [29] did not practically influence the smooth line 4 of the Neotectonic uplift. A moderate uplift of the Crimea began some 110 Ma after the end of the last intensive convergence in this area [59]. Strong convergence in the Carpathians terminated 11 Ma ago [8] and at that time the thrust belt was only a few hundred meters above sea-level. The present mountain range was formed mostly during the last 5 Ma (line 7). During this period of time the Carpathian foreland remained practically stable (line 8).

GENERAL FEATURES

We have considered several examples of the Neotectonic crustal uplift in regions with different age of the crystalline crust. This is Precambrian crust in southern Africa, Paleozoic crust in the Tien Shan and Mesozoic crust in the Verkhoyansk and Chersky ranges, and in the Crimea. The Greater and Lesser Caucasus and the Carpathians include the crust which formed both in the Mesozoic and Cenozoic. Together with a lot of the other data, these examples enable us to discuss general features of the Neotectonic crustal uplift and its possible driving mechanisms.

Global synchronicity

The Neotectonic uplift took place in two main stages. The first one occurred in the Late Oligocene or in the Early Miocene, as in southern Africa, the Tien Shan and Verkhoyansk and Chersky ranges. The same was typical of many other regions, e.g., the Tibetan Plateau [51], East Africa [58, 83], the Arabian Platform [16], Altay and Sayan mountains, the Transbaikalian area and Mongolia [103]. The major uplift began about 5 Ma ago; however, in different regions the time of its onset ranges from the Late Miocene to the Pleistocene. This has occurred in all the above mentioned regions, as well as on the Chinese Loess Plateau, in the Himalayas [102] and in the Andes [91].

Over most of continental areas the Neotectonic uplift was preceded by a period of relative stability, ~150 Ma long, which lasted from the Jurassic and until the Eocene–

Oligocene [70]. During this epoch a considerable uplift occurred in some regions from time to time. Such an uplift took place, e.g., before the break up of Gondwana on the future continental margins. During this long period of time, the rate of crustal uplift on the continents was, however, much lower on average than at Neotectonic time, especially compared with the last 5 Ma. In different regions the Neotectonic uplift took place at different time. The onset of the last and strongest event ranges from ~1.5 Ma ago to ~10 Ma ago. This time interval is, however, much smaller than the duration of the preceding epoch of relative stability (~150 Ma). This indicates a synchronicity of the Neotectonic uplift on a global scale in the first approximation.

The absence of significant horizontal deformations in most regions

The Neotectonic crustal uplift occurred on tectonic structures of various types: on the crystalline shields, in sedimentary basins and in the Phanerozoic fold belts. It took place on the crust of various age — from the Archean to Late Cenozoic. 72% of the total area of the continents are underlain by Precambrian crystalline crust [34]. Most of the Neotectonic uplift took place on the crust of this age, as a consequence. An intensive uplift (0.5-2 km) occurred, e.g., in southern and eastern Africa, on the Angola, Ethiopian and Somalian plateaus and on the East Siberian platform. A minor uplift took place on the Baltic, Australian, Canadian, Brazilian and Guiana shields.

Precambrian lithosphere, 550 Ma old, is characterized by a high stability. No significant stretching or shortening took place over a predominant part of Precambrian terrains during the Neotectonic uplift. Large horizontal deformations occurred only in some small regions. For example, the eastern and western branches of the East African rift zone were stretched by several tens of kilometres. At the same time, the inner part and the flanks of the East African Plateau, ~1000 km wide, which was uplifted by ~1 km, remained undeformed. Since most of the Neotectonic uplift has taken place on a stable Precambrian crust, it was generally associated with no significant lithospheric stretching or shortening.

The Neotectonic crustal uplift has widely occurred on the Paleozoic crust where strong shortening of the crust terminated long ago. In many cases, the Neotectonic uplift of such regions was associated with only slight horizontal deformations ≤1-2%. This is typical, e.g., of the Urals, Scandinavian Caledonides, Rheinisch and Bohemian massifs, Greenland, and the inner part of the Iberian Peninsula. Lithospheric shortening, stretching and strike-slip deformations of a low intensity widely occurred in Caledonides and Hercynides of Central Asia which was affected by collision of India with Eurasia. This took place in the Tien Shan, Altay, Sayan, Transbaikalian region, Mongolia and the Chinese Loess Plateau. In most regions horizontal deformations do not exceed 3-7%. For example, in north-eastern Mongolia, faults are nearly vertical and ensure almost no lithospheric shortening or stretching [56]. No significant shortening also occurred on the Tibetan Plateau at the time of the crustal uplift by several kilometres in the Pliocene and Pleistocene [51].

A similar situation is typical of most Mesozoic fold belts, e.g., of the Verkhoyansk and Chersky ranges and the Crimea. Their Neotectonic uplift was associated with only slight horizontal deformations. In fold belts which proceeded to evolve until the Late Cenozoic, strong convergence took place at the time of the Neotectonic uplift, as it happened, e.g., in the Himalayas, Pamirs, Caucasus, Kopet Dag and Atlas. Such fold belts, however, cover only a very small part of the continents altogether. Furthermore, as can be seen from the examples of the Caucasus and Kopet Dag, in some cases, short impulses of convergence did not strongly influence the uplift which evolved independently of strong episodic shortening of the crustal layer.

In some regions the Neotectonic uplift occurred synchronously with lithospheric stretching. This took place, e.g., on the Tibetan Plateau and in the Andes, in eastern Mongolia, in the East African rift zone and in the Basin and Range Province. Except the two wide terrains of the Basin and Range and Afar, intensive rifting took place in narrow zones which are located within strongly uplifted undeformed areas of a much larger width. Thus, in a predominant part of the areas of its occurrence, the Neotectonic uplift was not associated with significant convergence or lithospheric stretching.

Large gradients of vertical displacements

The Neotectonic uplift was associated with flexing of the lithosphere which formed steep slopes in many regions. The smoothly tilted wide interior plateau in southern Africa, which is ~1000 km wide, is surrounded by steep slopes which have been formed by the Neotectonic uplift; these slopes are up to ~1000 m high and only ~100-200 km wide (Fig. 2). Many mountains in the Phanerozoic fold belts were formed by predominantly a dome-like uplift. This was typical, e.g., of the Caucasus, Carpathians, Urals, Scandinavian Caledonides, and Verkhoyansk ranges. The western slope of the Verkhoyansk range in Fig. 4 is 1300 m high and 130 km wide. The northern slope of the Greater Caucasus is ~4 km high and 75-100 km wide. The northern slope of the Tibetan Plateau, 3-4 km high, is 100 km wide in many places.

Strong faulting has also occurred in some regions of the Neotectonic uplift (e.g., [95]). In most cases, the faults are steep and ensure no large horizontal deformations of wide areas; however, some of them are associated with large vertical displacements. Thus, the Tien Shan is bounded by a steep thrust fault in the south with vertical Neogene displacements up to 7-10 km. In the Chersky range in the right-hand side of Fig. 4 the isohypses of the uplift 600 m and 1000 m practically coincide in several places. This is associated with vertical displacements of the crustal blocks along the faults.

Association of the Neotectonic uplift with volcanism

An intensive volcanism took place in many regions of the Neotectonic uplift (however, not in all of them). The occurrence of volcanism is quite natural on active continental margins, e.g., in Japan, Camchatka and the Andes, on the young passive margins of the Red Sea, or potentially divergent boundaries such as the Basin and Range and East African rifts. Volcanism, however, has also occurred in many intraplate areas of the Neotectonic uplift — far from subduction zones and without strong lithospheric stretching. Thus, a strong increase in the volcanic activity took place in the Neogene on the crystalline shield of North Africa: the Ahagger, Tibesti and Darfur plateaus, the Chad Basin and the Cameroon Chain [22, 99]. Intensive volcanism, occurred in the Pliocene and Pleistocene on the Tibetan Plateau [40, 75], in the Caledonides and Hercynides of Mongolia and in the East Sayan range to the west of lake Baikal [52]. Late Cenozoic and Quaternary volcanism took place in the Hercynides of the Massif Central in France [33] and the Rheinisch Massif in Germany [18, 76]. The beginning of Neotectonic uplift in the Atlas and Tien Shan in the Oligocene was associated with slight volcanism [55, 70].

Asthenospheric upwelling under some regions of the Neotectonic crustal uplift
A strong rise of the top of the asthenospheric layer is observed under some continental regions where a large Neotectonic uplift has taken place. For example, on south-eastern margin of the East Siberian Platform, which is followed by the Baikal rift zone, the top of the high electrical conductivity layer rises from a depth of 180-200 km under the platform to 75-90 km under the lake (Fig. 6) [43]. The strong increase in the electrical conductivity indicates partial melting and high temperature in the mantle. According to the seismic data, an upwarp of the low-velocity layer by ~50 km takes place in the same region which extends further to the south-east into the Transbaikalian region [28].

According to the seismic data, a rise of the asthenospheric layer has been also found beneath the Atlas mountains [86] where a strong crustal uplift took place in the Neogene [27]. In the Eastern Tien Shan the asthenospheric upwelling reaches the base of the crust with a poorly expressed Moho boundary [49, 81]. In this region, the character of seismic anisotropy in the mantle indicates recent replacement by hot asthenosphere of mantle lithosphere which formed in the Paleozoic and Early Mesozoic due to a small scale convection [61]. According to the geothermic data, the crust of the Lesser Caucasus where a strong uplift occurred in the Pliocene to Pleistocene (Fig. 5) is underlain by a hot mantle with a temperature T ~1000-1200 C [88]. No asthenospheric upwelling has been, however, found using the seismic data beneath cool cratonic areas of the Neotectonic uplift. In, for example, southern Africa high velocities in the mantle are observed down to the transitional layer (400 km) [96, 97] and only a thin low-velocity layer probably exists just above the transitional layer.

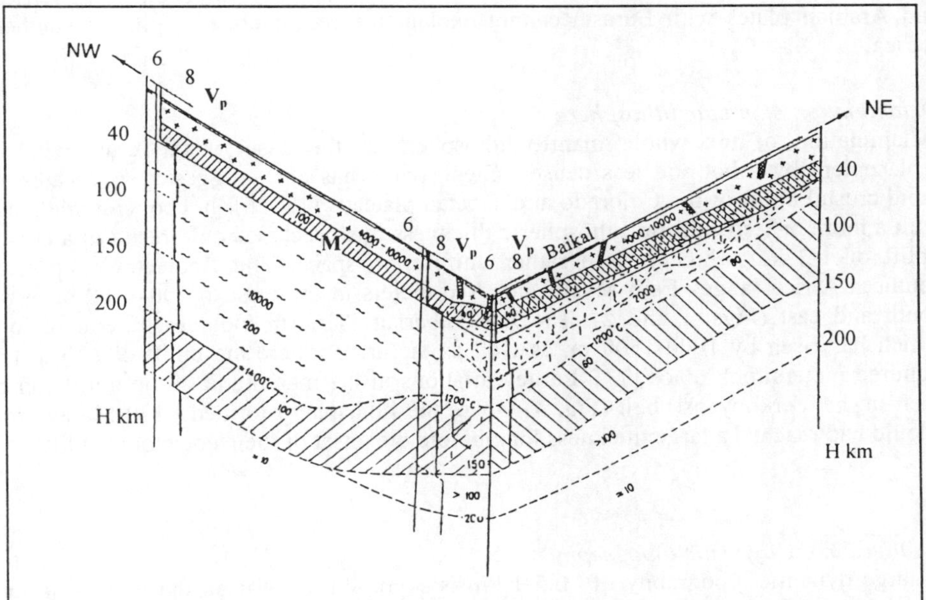

Figure 6. Electrical resistivity, mantle temperature and P-wave velocities (km/s) under the Siberian Platform and the Baikal rift zone according to magnetotelluric and deep seismic sounding data (modified after [43]).

POSSIBLE DRIVING MECHANISMS

The isostatic response to shortening and thickening of the crust
This is a classical explanation for the formation of high mountains in fold belts [e.g., 60]. Most of the Neotectonic uplift, however, has taken place on the Precambrian, Paleozoic and Mesozoic crystalline crust where intensive shortening terminated long before the onset of the uplift. Fold belts where crustal shortening proceeded into the Late Cenozoic cover only a very small part of the continental areas. In some of them, e.g., in the Carpathians the present mountains formed after the end of convergence.

The formation of a high topography in Central Asia is commonly attributed to collision of the Indian plate with Eurasia [63] which began ~50 Ma ago. The major crustal uplift in the Himalayas, on the Tibetan and Chinese Loess plateaus, in the Pamirs and Tien Shan, in the Altay and Sayan mountains, in the Transbaikalian region and Mongolia took place during the last 3-5 Ma, i.e., after 90% of collision were over. In most regions it was associated with horizontal crustal deformations of no more than a few per cent. Shortening of the crust in the Tien Shan in the Pliocene and Pleistocene by ~5% on average was sufficient to produce the isostatic crustal uplift by 300 m. In fact, during this period of time the Tien Shan has risen by about 3 km. The uplift of the Tibetan Plateau by 3.5-4 km during the last 3.5 Ma has occurred without significant shortening of the crust [51]. Collision of the Indian and Eurasian plates could not also produce a strong uplift far from the collisional boundary in East Siberia and north-eastern Asia. In particular, it could in no way affect the formation of the Verkhoyansk and Chersky ranges which are 4500-5000 km from this boundary. Similarly, collision of the African and Arabian plates with Eurasia cannot explain the recent crustal uplift in southern Africa.

Delamination of mantle lithosphere
Delamination of the whole mantle lithosphere at the base of the crust with its replacement by a hot and less dense asthenosphere has been suggested as a cause of rapid crustal uplift of the Colorado and Tibetan plateaus [13, 26]. In cool cratonic areas with a more or less uniform lithospheric thickness this process should result in a crustal uplift of 1-2 km which is also rather uniform in space. The Neotectonic uplift of southern Africa ranges from several hundred meters in the west to 700-1100 m in the south and east (Figs. 1 and 2). The East Siberian Platform includes several regions which has risen by 1000-1500 m, as well as the areas where the uplift of only a few hundred meters took place [71]. Large variations in the magnitude of the uplift can be seen in the Verkhoyansk belt (Fig. 4) where the lithosphere formed ~120 Ma ago and should had reached a large thickness long before the onset of the Neotectonic uplift.

Changes in the dynamic topography
A large dynamic topography of ~0.5-1 km is supposed to exist on the Earth's surface which is produced by convective flows in the mantle [e.g., 36]. Changes in this topography can result in vertical crustal movements. Such an explanation has been suggested, e.g., for the Neotectonic uplift of East Africa [25]. The width of the areas

with a constant sign of this topography should be several thousand kilometres or more with very smooth lateral variations. As was mentioned above, in many regions the magnitude of the Neotectonic uplift varies by 0.5-4 km at short distances 50-200 km. These narrow-wavelength variations cannot be produced by changes in the dynamic topography which have a smooth profile.

The isostatic response to denudation

The Neotectonic crustal uplift has been explained by the isostatic response to rapid denudation of a high pre-existing topography, which began a few million years ago due to the climatic changes [62]. As follows from the analysis of topography of the Alps, Himalayas, Karakoram and Andes, this effect could actually increase the altitude of the highest peaks by no more than one quarter of their present height [32]. The effect must be considerably smaller for most of Africa and for the inner part of the Tibetan Plateau which are dissected only slightly. A shallow sea covered the Central Asia in the Eocene and most of the Greater Caucasus at the start of the Late Miocene. High mountains in these regions could be formed solely by a tectonic crustal uplift. Similarly, some regions in the East Siberian Platform which are predominantly covered by marine deposits have risen by up to 1-1.5 km during Neotectonic time.

The crustal uplift on passive continental margins has been explained by a flexural isostatic response to differential denudation after rifting and continental break up [30, 31]. In particular, this mechanism ensured a large crustal uplift after the break up of Gondwana in the Mesozoic. A sophisticated model has been formulated which describes the crustal uplift and the formation of escarpments after continental rifting and the following denudation of the topography [47, 48]. This process was active after the continental break up, with the intensity rapidly decreasing with time. In most regions rapid denudation ceased long before the onset of the Neotectonic uplift. Thus, erosion on the margins of southern Africa was very intensive after the continental break up in the Early Cretaceous; however, it became slow in the Late Cretaceous-Early Miocene [39]. Rapid denudation of these regions which had ceased ~ 100 Ma ago could not ensure the Neotectonic uplift which began 18 Ma ago. Furthermore, in the Pliocene and Pleistocene a large uplift with the magnitude increasing eastward took place on the interior plateau (Figs. 1 and 2). As pointed out by Hartnady and Partridge [39], the uplift in this area could not be significantly influenced by denudation of distant coastal regions.

Flexing and heating of the lithosphere after its rifting

Mechanical unloading of the lithosphere due to its stretching has been proposed as a cause of flexural uplift on rift shoulders [e.g., 20, 98]. Another mechanism suggested for the uplift near to divergent plate boundaries is lateral heat migration and secondary convection [90]. Over most of the areas of its occurrence the Neotectonic uplift took place far from strongly rifted basins. The uplift on rift shoulders took recently place only in some places, e.g., on the margins of the Red Sea or of the East African rifts. Intensive rifting and block tilting on the margins of the Red Sea took place in the Oligocene [14, 41]. However, this was associated with no strong crustal uplift. At that time the Arabian rift shoulder remained a few hundred meters above sea-level, and the Cretaceous-Eocene planation surface which had formed at a low altitude still existed there [16].

The major uplift of the Arabian Platform which was associated with the formation of coarse deposits, up to ~2000 m thick, with a large volume of conglomerates began later: ~5 Ma ago in the north and ~3 Ma ago in the south. This has risen the platform margin to an altitude of ~2 km. The African margin of the Red Sea is still low. Thus, there was no significant flexural response to rifting on the Arabian margin of the Red Sea. This can probably be explained by a strong synchronous decrease in the effective elastic thickness of the lithosphere [8, 10, 11]. The uplift due to secondary convection and lateral heat migration should slow down with time [90]. These phenomena cannot explain a strong uplift of the Arabian Platform which began ~20 Ma after rifting.

The uplift from rising megaplumes

Megaplumes of hot and low-density material which are rising from the mantle or spreading along the base of the lithosphere can, in principle, produce a large uplift on the surface [35]. To support an uplift of ≥1 km, the thickness of a hot layer should be ~150-200 km. In some regions of a pronounced Neotectonic uplift, e.g., in southern Africa and East Siberia, according to high seismic velocities, the mantle is cool down to a large depth of 300-400 km [78, 93]. Moreover, a megaplume, ~1000 km in diameter should produce an uplift with a smooth profile and the maximum near the centre of the megaplume. As mentioned above, in fact, strong lateral changes in the magnitude of the uplift often occur in regions which are only ~50-100 km wide. The magnitude of the Neotectonic uplift in southern Africa increases not toward the inner part of the interior plateau, but toward its eastern margin (Fig. 2).

The uplift from changes in the force which is acting along the lithospheric layer

The crustal uplift and subsidence can arise from time variations in a force acting along the lithospheric layer [21]. Typical vertical displacements under such circumstances are estimated to be ~100 m and such effects would be inversely proportional to the square width of the area [2]. In areas, ~1000 km wide, vertical displacements of the lithosphere would be negligible. Furthermore, large deflections of this layer from isostasy would give rise to large isostatic anomalies which are not usually observed in areas of the Neotectonic uplift [1, 38, 67].

EROSION OF MANTLE LITHOSPHERE A POSSIBLE MAIN CAUSE OF THE NEOTECTONIC UPLIFT

A density decrease in the lithosphere as a general cause of the uplift

As follows from the above considerations, many different mechanisms have been proposed to explain the Neotectonic crustal uplift in different regions. Each of them can be active under certain specific conditions. For example, the Neotectonic uplift from shortening of the crust could have occurred only in regions of strong Late Cenozoic convergence. The uplift on passive continental margins can be discussed for a few places of recent continental break up like the margins of the Red Sea or the Gulf of California. Recent uplift on rift shoulders could have occurred only on the margins of young rifts. Delamination of a thick mantle lithosphere can happen in any place and at any time; however, it will produce the uplift which is uniform in space. In this approach, the Neotectonic uplift over about one half of the continental areas should be attributed to a synchronous manifestation of different independent mechanisms. Then a question arises, why all these mechanisms were much less active altogether during the preceding ~150 Ma and suddenly began to operate ~30 Ma ago and mostly 3-5 Ma ago, while the character of plate motions did not considerably change as compared to the rest of the Cenozoic and Mesozoic.

A synchronicity of the Neotectonic crustal uplift in the areas which are far apart indicates that in most cases it should be produced by a common basic mechanism. In a predominant part of the cases, the Neotectonic uplift took place far from plate boundaries on the Precambrian and, up to a smaller extent, on the Paleozoic and Mesozoic crust where lithospheric shortening or stretching did not exceed a few per cent. In most of the uplifted regions the lithosphere is close to isostasy [1, 38, 67]. Together with high gradients of the Neotectonic uplift which are typical of many places, this indicates that the uplift resulted from a density decrease at shallow depth — in the lithospheric layer.

The uplift in cool areas as a consequence of rapid asthenospheric upwelling
A density decrease in the lithospheric layer can take place in the crust or in mantle lithosphere. In mafic rocks of the lower crust (in regions where it has a mafic composition) a considerable density decrease can result from a phase transformation of garnet granulites into a less dense gabbro [4, 6, 8]; however, this process occurs only at a temperature $T \geq 800$ C. Another possibility is delamination of dense eclogites with their sinking into the mantle [4]. This requires upwelling of the asthenosphere to the base of the crust with even stronger heating of the lower crust. Many regions of the Neotectonic uplift are actually cool [72, 88] with a top of the asthenosphere far below the Moho. This is typical, e.g., of southern Africa, the Victoria Plateau in East Africa and of East Siberia. The crustal uplift in such areas could have occurred only due to a density decrease below the crust — due to a replacement of the lower part of mantle lithosphere by a less dense asthenosphere.

Asthenospheric upwelling from conductive heating of a thick layer in mantle lithosphere (~100 km) takes tens of million years. The Neotectonic uplift in the Pliocene and Pleistocene was much more rapid (~5 Ma). In cool areas a thick mantle layer remains preserved in the lithosphere which indicates that a complete delamination of mantle lithosphere did not take place. In such areas, the Neotectonic uplift required a rapid erosion of mantle lithosphere at the lithosphere-asthenosphere boundary which has been proposed by many authors [e.g., 3, 23, 64, 80, 87, 100]. There are no direct geophysical data which could reveal the existence or non-existence of a significant erosion of the lithosphere in oceanic areas. As was mentioned above, a rise of the asthenospheric layer of ~50-100 km has been, however, found in some areas of the Neotectonic uplift on the continents.

Consider, e.g., the asthenospheric upwelling beneath the margin of the Siberian Platform (Fig. 6). Apart of lake Baikal, a normal heat flow is observed in the area. This indicates that the rise of the top of the asthenosphere has formed during the last ≤10 Ma, i.e., the asthenospheric upwelling has taken place quite recently. The crustal uplift by 0.5-1 km occurred in the same area in the Pliocene and Pleistocene [103]. Upwelling of hot asthenosphere by ~100 km is sufficient to produce the observed uplift. Except in lake Baikal, the uplift was associated with no significant stretching of the lithosphere. In lake Baikal extension of the crust is ≤5-10 km [7]. Thus, the average extension of the uplifted area does not exceed a few per cent, while the lithosphere is thinned by about two times. This shows that the asthenospheric upwelling was of an active character.

Asthenospheric upwelling as a possible consequence of infiltration of volatiles into mantle lithosphere from the anomalous mantle

Mantle lithosphere is cooler and denser than the asthenosphere. This can, in principle, result in convective replacement of mantle lithosphere by a less dense asthenosphere with the associated crustal uplift. Before the onset of the Neotectonic uplift most of continental areas were near to sea-level during a period of time, ~150 Ma long [70]. This indicates that under normal conditions mantle lithosphere has a high viscosity, and the characteristic time of convective instability in the system "mantle lithosphere-asthenosphere" is larger than the time of existence of mantle lithosphere. Tectonic deformations can enhance the instability; however, no significant deformations preceded the Neotectonic uplift in most regions, especially on the Precambrian terrains. Hence a strong decrease in the viscosity of mantle lithosphere was necessary to ensure its rapid replacement by the asthenosphere.

Two main possibilities can be discussed. Firstly, mantle viscosity h_m can be reduced by heating which decreases h_m by one order of magnitude per each 100 C [46, 92]. For a strong viscosity drop, mantle lithosphere should be heated up to a temperature T which is close to that of the asthenosphere ($T \approx T_a$). This can be ensured by upwelling to the base of the lithosphere of an anomalously heated mantle material — the anomalous mantle with the temperature $T_{am} > T_a$, which can occur, e.g., in hot plumes. The temperature of mantle lithosphere decreases upwards. To heat a layer of mantle lithosphere of a thickness Δh_{ml} ~50-100 km up to $T \approx T_a$, the thickness of a layer of the anomalously heated mantle under the lithosphere (h_{am}) should be several times larger than Δh_{ml}. In many regions of intensive Neotectonic uplift the upper mantle above the transitional layer (down to 400 km) is, however, cool. This is typical of, for example, southern Africa and East Siberia [78, 96, 97]. In such regions, a partial or complete collapse of mantle lithosphere due to its heating from the anomalous mantle is improbable.

The second possibility is that the anomalous mantle which ascends to the base of the lithosphere can be fluid-containing. In wet rocks the viscosity is several orders of magnitude lower than in dry ones. Infiltration of volatiles from the anomalous mantle into mantle lithosphere can result in a strong viscosity drop in this layer or in its lower part. In some regions such a viscosity drop had taken place in the whole lithospheric layer [8, 10, 11]. That part of mantle lithosphere where the viscosity is reduced, will delaminate and sink into the asthenosphere. In this case, a large asthenospheric upwelling can develop even after emergence to the lithosphere of a comparatively small volume of fluid-containing anomalous mantle. Infiltration of volatiles ensures a collapse of mantle lithosphere which can be then replaced mostly by normal asthenosphere. Depending on the amount of volatiles in the anomalous mantle, a partial or complete collapse of mantle lithosphere will occur. It should be noted that infiltration of volatiles has been earlier suggested as a possible cause of the Neotectonic uplift by King [44]. A synchronous uplift in many areas which are thousands of kilometres apart indicates the occurrence of this phenomenon on a global scale.

The magnitude of the uplift due to the asthenospheric upwelling

Suppose that asthenospheric upwelling takes place due to a replacement of mantle lithosphere by normal asthenosphere with the temperature T_a. Assume that at the initial moment the temperature in the lithosphere increases linearly with the depth reaching T_a at its base. Let the initial thickness of mantle lithosphere be h_m and the Moho temperature — T_M. Designate thermal expansivity of the mantle by α. Consider a wide area which is close to local isostasy. Then a rapid replacement by the asthenosphere of the lower part of mantle lithosphere of the thickness Δh_{ml} results in the crustal uplift:

$$\Delta\zeta \approx \alpha(T_a - T_M)\,\Delta h_{ml}^2/2h_{ml}. \qquad (2)$$

This quantity increases as a square of the magnitude of the upwelling Δh_{ml}. Let us take $\alpha = 3\cdot10^{-5}$, $T_a = 1300$ C, $T_M = 500$ C, and the initial thickness of mantle lithosphere in a cool area $h_{ml} = 150$ km. Then the asthenospheric upwelling by $\Delta h_{ml} = 50$ km and $\Delta h_{ml} = 100$ km will result in the uplift by $\Delta\zeta = 200$ m and $\Delta\zeta = 800$ m, respectively. Upwelling of the asthenosphere to the base of the crust will produce the uplift of 1800 m.

Spreading of the anomalous mantle along the base of the lithosphere.

The Neotectonic crustal uplift took place both in regions where the upper mantle above the transitional layer is hot, as in western North America or north-eastern Africa, and cool, as in southern Africa and East Siberia. In cool areas, only a thin layer of hot anomalous mantle can exist at the base of the lithosphere and its ascent from the underlying mantle is improbable. This hot and low-viscosity material could have been transported into cool regions from the adjoining hot areas by a rapid lateral flow along the base of the lithosphere (Fig. 7). Interaction of the flow of the anomalous mantle

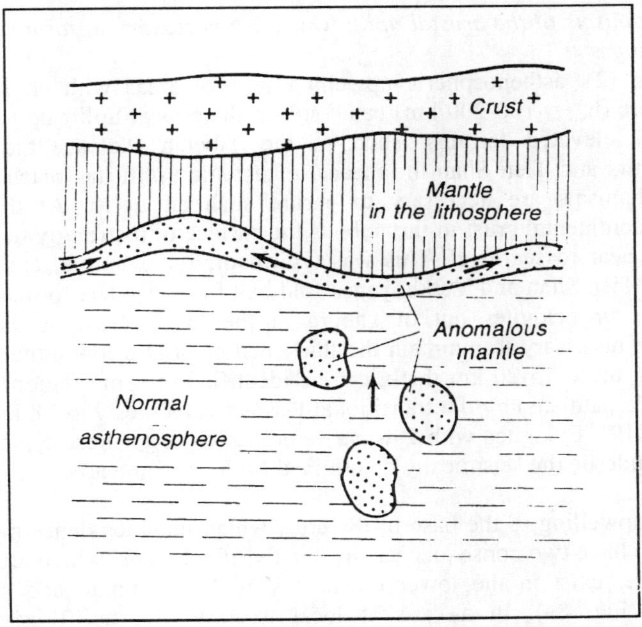

Figure 7. The flow of high temperature and low-viscosity anomalous mantle along the base of the litho-sphere with filling of the trap at this boundary (modified after [3]).

along the base of the lithosphere with lateral variations of the lithospheric thickness can explain a large lateral nonuniformity of the Neotectonic uplift. The lithospheric thickness strongly varies laterally (Fig. 7). Volumes of the anomalous mantle well up from a deep mantle, reach the base of the lithosphere and spread along this boundary. A region with a reduced lithospheric thickness represents a trap for the anomalous mantle. Accumulation of the anomalous mantle in the trap produces the crustal uplift at the surface which can increase due to erosion of mantle lithosphere by the anomalous mantle. The hot material becomes trapped in the lithosphere and cannot spread rapidly aside. This permits to keep high gradients of topography which are typical of many regions of the Neotectonic uplift.

The conditions that necessary for a rapid flow of the anomalous mantle along the base of the lithosphere have been considered in [3, 9]. To allow a lateral spreading of the anomalous mantle layer, ~20 km thick, along this boundary for a distance of ~1000 km during the period of time, ~1 Ma long, the viscosity of normal asthenosphere should be $\eta_a \leq 3\cdot10^{18}$ Pa s. This is much lower than the estimates of $\eta_a \sim 10^{20}$ Pa s in convection models which are based on the data on the geoid following [36]. The above low value of η_a is, however, close to the estimates of asthenospheric viscosity in continental areas $\eta_a \sim 10^{18}$-10^{19} Pa s which are based on the data on postglacial rebound (see, e.g., [3, 65]). To ensure a rapid flow of the material within the layer of the anomalous mantle, its viscosity should be $\eta_{am} \leq 10^{17}$ Pa s which is ~30 times smaller than $\eta_a \sim 3\cdot10^{18}$ Pa s. This condition can be easily fulfilled, if the anomalous mantle is fluid-containing, especially if it is hotter than normal asthenosphere.

Additional mechanisms of the crustal uplift which are necessary to form high plateaus and mountain ranges

As follows from (2), asthenospheric upwelling in cool areas with an initially thick mantle lithosphere (h_{ml} = 100-200 km) can result in the crustal uplift up to 1.2-2.4 km. In some highly elevated terrains, such as the Tibetan Plateau, the Himalayas, Karakoram, Pamirs and Tien Shan the mean Neotectonic uplift has reached 3.5-4 km. Additional mechanisms are necessary to ensure such an uplift. At the epochs of convergence of continental crust in thrust belts the surface of a strongly thickened crust usually remains near to sea-level. This was typical of, for example, the Carpathians, Urals, Paleozoic Tien Shan and Verkhoyansk fold belt [4, 6, 8]. The formation of dense garnet granulites or eclogites out of gabbro in the lower crust at the epochs of convergence was necessary to maintain the thickened crust at a low altitude. Layers of garnet granulites, up to 15-20 km thick, can be identified in some regions (e.g., in the Urals and East Carpathians) by the transitional P-wave velocities 7.4-7.8 km/s in a cool lower crust [12, 19]. Eclogites with P-wave velocities close to those typical of mantle peridotites can underlie the seismically determined Moho in some areas.

Asthenospheric upwelling to the base of the crust which includes dense mafic rocks in its lower part can have two consequences [4]. Firstly, if eclogites, which are denser than the asthenosphere, exist in the lower crust, they will delaminate and sink into the asthenosphere (Fig. 8a). In a case of local isostasy, the uplift associated with delamination of a layer of eclogite with the density ρ_e and thickness h_e is:

$$\Delta\zeta_e = [(\rho_e - \rho_a)/\rho_a]h_e, \tag{3}$$

where ρ_a is the density of the asthenosphere. Taking ρ_e = 3500 kg/m^3, ρ_a = 3220 kg/m^3 and h_e = 20 km, we find: $\Delta\zeta_e$ = 1.7 km. Secondly, the lower crust can include a layer of garnet granulites which are less dense than the asthenosphere and cannot sink into it. Under a contact with the asthenosphere, garnet granulites will be heated with a retrogression back to gabbro. Expansion in a layer of garnet granulites with the density ρ_{gl} and the initial thickness h_{gl} will produce the isostatic uplift:

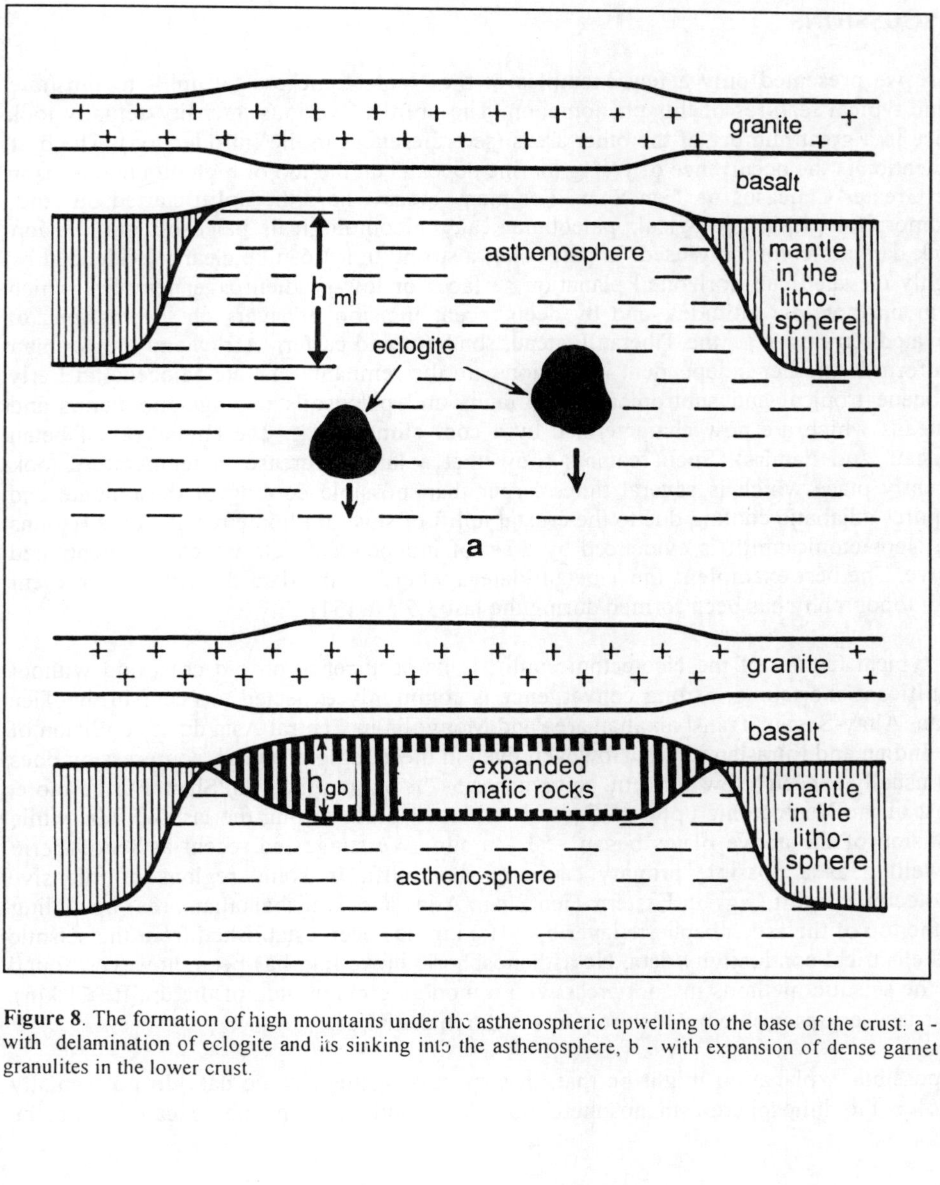

Figure 8. The formation of high mountains under the asthenospheric upwelling to the base of the crust: a - with delamination of eclogite and its sinking into the asthenosphere, b - with expansion of dense garnet granulites in the lower crust.

$$\Delta\zeta_{gl} = [(\rho_{gl} - \rho_{gb})/\rho_{gb}]h_{gl}, \quad (4)$$

where ρ_{gb} is the density of gabbro. Taking ρ_{gl} = 3200 kg/m^3, ρ_{gb} = 2930 kg/m^3 and h_{gl} = 20 km, we have: $\Delta\zeta_{gl}$ = 1.8 km. Together with the crustal uplift from the asthenospheric upwelling, these two mechanisms can ensure the maximum observed magnitudes of the Neotectonic crustal uplift of 3.5-4 km. At the initial altitude of the crustal surface 0.5-1 km, this will form plateaus and mountain ranges with a mean altitude up to 4.5-5 km. Individual peaks can rise higher, e.g., due to dissection of the topography as suggested by Molnar and England [62].

DISCUSSIONS

Here we presented only a few examples of the Neotectonic crustal uplift to illustrate some typical features of this phenomenon. The uplift of various intensity actually took place in a great number of the other areas (see references in the Introduction). The best indication is the occurrence of young marine deposits on the top of high mountains, as in the Greater Caucasus or Tien Shan. For many areas where the uplift started on land, geomorphological, geological, paleobotanical, paleontological, paleosoil and fission track data are commonly used. In particular, a strong uplift can be clearly evidenced by highly elevated sub horizontal planation surfaces or low-gradient river terraces, which form only at low altitudes, and by deep recent incision of rivers on the margins of elevated terrains (e.g., the Tibetan Plateau, southern and eastern Africa and the Arabian Platform). Another independent indications are the remnants of Late Miocene an Early Pliocene tropical and subtropical flora, fauna and paleosoils on high mountains and plateaus which are now characterized by a cool climate (e.g., the Himalayas, Tibetan Plateau, and Pamirs). Such features show that a large decrease in temperature took recently place which is several times larger than possible cooling of the climate and requires adiabatic cooling due to the crustal uplift of several kilometres. In some regions the Neotectonic uplift is evidenced by a set of independent data which are mentioned above. The best example is the Tibetan Plateau where all the data show that the present high topography has been formed during the last 3.5 Ma [51].

A typical feature of the Neotectonic uplift is its occurrence on old crust and without significant shortening. Strong convergence is commonly expected to occur in the Tien Shan, Altay-Sayan, Transbaikalian area and Mongolia in Central Asia due to collision of the Indian and Eurasian plates. However, even in these regions lateral compression does not usually exceed a few per cent, and reaches 5-7% only in the Tien Shan. Furthermore, most of the Neotectonic uplift in Central Asia took place during the last 3-5 Ma, while collision of the above plates began ~50 Ma ago. We suggested recent asthenospheric upwelling as a possible primary cause of the uplift. In some regions of intensive Neotectonic uplift (e.g., in Eastern Tien Shan, Atlas and Transbaikalian area) upwelling of the top of the asthenospheric layer by ~100 km has been established from the seismic and electrical conductivity data. No asthenospheric upwelling has been, however, found by the seismic methods in cool areas with a moderate magnitude of the uplift ($\leq$1 km). This is a serious problem which requires additional studies.

A possible explanation might be that, in some cases, the seismic data do not actually resolve the lithosphere-asthenosphere boundary, and the lithosphere can in fact be thinner than the high velocity lid. Thus, in East Siberia the thickness of this layer is ≥300 km [77], while the top of the high conductivity layer under the southern part of the area is at a depth of ~200 km (Fig. 6). A large increase in the electrical conductivity indicates partial melting of the mantle which is associated with a drastic decrease in its viscosity. Hence the lithospheric thickness beneath southern East Siberia cannot exceed ~200 km. Thus, even the maximum possible thickness of the lithosphere in this region (~200 km) is ~100 km larger than the thickness of the high velocity lid (~300 km). A detailed analysis of the seismic and electrical conductivity data on different regions of the Neotectonic uplift is necessary to solve the problem.

Another question is a cause of rapid collapse of mantle lithosphere, partial or complete, which has allowed the asthenospheric upwelling under some areas of the Neotectonic uplift. A tentative suggestion is that this collapse could be associated with emergence of fluid-containing anomalous mantle to the base of the lithosphere and infiltration of volatiles into mantle lithosphere. This is in agreement with volcanism which is typical of many regions of the Neotectonic uplift; however, a detailed analysis of the geochemical data is necessary to check this possibility. The deep seated origin of the anomalous mantle requires detailed tomographic studies. A cause of synchronous emergence of the anomalous mantle beneath different continents should also be explained. Thus, a global occurrence of the Neotectonic crustal uplift poses many new problems associated with its physical mechanism.

CONCLUSIONS

As follows from a large volume of data, a strong crustal uplift began in many continental areas 20-30 Ma ago after a period of relative stability, ~150 Ma long. In most regions, this Neotectonic uplift took place in two impulses. The first one occurred in the Late Oligocene or Early Miocene after a period of relative stability, ~150 Ma long. The second, and in most regions the major one began 3-5 Ma ago. Its magnitude ranges from a few hundred meters to a few kilometres. The uplift has developed with a strong acceleration in time and formed a predominant part of the present high mountains and plateaus, such as mountain ranges in the Alpine and Verkhoyansk fold belts, the Tien Shan, Altay-Sayan, the South and East African plateaus. Most of the uplift has occurred on the Precambrian, Paleozoic and Mesozoic crust where intensive shortening terminated long ago. This precludes a possible role of a classical mechanism of the isostatic response to crustal thickening.

While the phenomenon of a rapid Neotectonic uplift is well documented by a large volume of data, we can only speculate about its origin. The uplift occurred in many wide intraplate areas far from active plate boundaries and without a strong disturbance of isostasy (in, e.g., East Siberia or southern Africa). In such areas, a significant role of plate-boundary-related forces and processes is precluded and a density decrease in the crust and/or mantle is necessary to ensure the uplift. Changes in the dynamic topography or megaplumes rising in the mantle should produce wide and smooth swells on the surface. In fact, the magnitude of the Neotectonic uplift varies by 0.5-4 km in many regions which are only ~100 km wide. This is an indication on a shallow depth of compensation.

A thick lithospheric layer is still preserved in most regions of the Neotectonic uplift which is incompatible with delamination of the whole mantle lithosphere at the Moho boundary. A partial erosion of this layer at the lithosphere-asthenosphere boundary can be tentatively suggested as a possible cause of the uplift. The asthenospheric upwelling which has been produced by this process exists beneath, for example, the Baikal and Transbaikalian regions. Before the onset of the Neotectonic uplift, in most continental areas mantle lithosphere remained stable during a period of time, 150 Ma long. Its recent collapse in many regions could probably be associated with emergence to the base of this layer of a fluid-containing anomalous mantle in a number of places with weakening of mantle lithosphere by infiltration of volatiles.

The occurrence of the Neotectonic uplift in areas with both hot and cool asthenosphere can be an indication on a rapid spreading of the anomalous mantle along the base of the lithosphere from the former areas into the latter ones. In the initially cool cratonic areas asthenospheric upwelling to the base of the crust can ensure the crustal uplift up to 1.5-2.5 km. The maximum observed magnitude of the uplift (3.5-4 km) requires additional processes. These might be delamination of dense eclogites or expansion of heated dense garnet granulites in the lower crust under a contact with the asthenospheric upwelling.

To produce the Neotectonic uplift erosion of several 10% of the total volume of continental mantle lithosphere was necessary. According to the scale of this erosion and that of the Neotectonic uplift these are the most powerful phenomena which occur in continental lithosphere.

Acknowledgements

In our analysis we used a large volume of data which had been provided to us by many scientists. Among them are E. Banda, M.A. Baer, E. Briem, G. Büchel, H. Faure, M. Fernandez, J.F. Gellert, R. Guiraud, I. Hessland, O. Kandler, V.E. Khain, A. Kröner, N.V. Lukina, V.I. Makarov, E.E. Milanovsky, N.-A. Mörner, A.A. Neumark, C.D. Ollier, T.C. Partridge, C. Prodehl, V.A. Selivanov, J. Sougy, M.R. Strecker, V.G. Trifonov, M. Villeneuve and L.P. Vinnik. We also got a lot of help from I. Fetzer.

REFERENCES

1. M.E. Artemjev, M.K. Kaban, V.A. Kucherinenko, G.V. Demyanov and V.A. Taranov. Subcrustal density inhomogeneities of Northern Eurasia as derived from the gravity data and isostatic models of the lithosphere, *Tectonophysics* **240**, 249-280 (1994).
2. E.V. Artyushkov. Can the Earth's crust be in a state of isostasy, *Jour. Geophys. Res.* **79**, 741-752 (1974).
3. E.V. Artyushkov. *Geodynamics*. Elsevier, Amsterdam (1983).
4. E.V. Artyushkov. *Physical Tectonics*. Nauka, Moscow (1993) (in Russian).
5. E.V. Artyushkov and M.A. Baer. Mechanism of formation of deep basins on continental crust in the Verkhoyansk fold belt: miogeosynclines and cratonic basins, *Tectonophysics* **122**, 217-245 (1986).
6. E.V. Artyushkov and M.A. Baer. Mechanism of formation of fold belts: the Alpine-Carpathian Region. In: *The Origin of Arcs*. F.-C. Wezel (Ed.). pp. 191-232. Elsevier, Amsterdam (1986).
7. E.V. Artyushkov, M.A. Baer, F.A. Letnikov and V.V. Ruzhich. On the mechanism of graben formation, *Tectonophysics* **197**, 99-115 (1991).
8. E.V. Artyushkov, M.A. Baer and N.-A. Mörner. The East Carpathians; indications for phase transitions, lithospheric failure and decoupled evolution of thrust belt and its foreland, *Tectonophysics* **262**, 101-132 (1996).

9. E.V. Artyushkov and A.W. Hofmann. Neotectonic uplift and magmatism on the continents and mantle convection. The case of southern Africa, *Surv. in Geoph.*, subm.
10. E.V. Artyushkov and N.-A. Mörner. Steep basement slopes in sedimentary basins: an indication on lithospheric failure, *Terra Nova* subm.
11. E.V. Artyushkov and N.-A. Mörner. Loss of lithospheric rigidity in sedimentary basins which is independent on plate collision and lithospheric stretching, *Geophys. Jour. Int.* subm.
12. V.V. Beloussov, N.I. Pavlenkova and G.N. Kvyatkovskaya (Eds.). *Deep structure of the USSR territory*. Nauka, Moscow (1991) (in Russian).
13. P. Bird. Continental delamination and the Colorado Plateau, *Jour. Geophys. Res.* **87**, 7561-7571 (1979).
14. R.G. Bohannon, C.W. Naeser, D.L. Schmidt and R.A. Zimmermann. The timing of uplift, volcanism, and rifting peripheral to the Red Sea: A case for passive rifting ? *Jour. Geophys. Res.* **94**, 1683-1701 (1989).
15. G.C. Bond. Evidence for some uplifts of large magnitude in continental platforms, *Tectonophysics* **61**, 285-305 (1979).
16. E. Briem. Die morphologische und tectonische Entwicklung des Roten Meer-Grabens, *Z. Geomorph. N.F.* **33**, 458-498 (1989).
17. *Bulletin of the INQUA Neotectonic Commission*. N.-A. Mörner (Ed.), Stockholm (1978-1996).
18. G.V. Büchel, V. Lorenz, H.-U. Schmincke, and B. Zimanovski. *Quartäre Vulkanfelder der Eifel, Fortschr. Miner., Bd.* **64**, *Beiheft* **2**, 97-141, Stuttgart (1986).
19. A.V. Chekunov (Ed.). *Geophysical studies of the lithosphere*. Naukova Dumka, Kiev (1993) (in Russian).
20. J. Cherie, F. Lucazeau, M. Daigniers and J.P. Vilotte. Large uplift of rift flanks: A genetic link with lithospheric rigidity, *Earth Planet. Sci. Lett.* **112**, 195-211 (1992).
21. S. Cloetingh, H. McQueen and K. Lambeck, On a tectonic mechanism for regional sealevel variations, *Earth Planet. Sci. Lett., 75,* 157-166 (1985).
22. J.M. Dantria and A. Lesquer. An example of the relationship between rift and dome: recent geodynamic evolution of the Hoggar swell and of its nearby regions (Central Sahara, Southern Algeria and eastern Niger), *Tectonophysics* **163**, 45-61 (1989).
23. G.D. Davies. Thermomechanical erosion of the lithosphere by mantle plumes, *Jour. Geophys. Res.* **99**, 15709-15722 (1994).
24. R.V. Dingle. Continental margin subsidence: A comparison between the east and west coasts of Africa. In: *Dynamics of Passive Margins*. R.A. Scrutton (Ed.). pp. 59-71. *Geodynamic Series* **6**, Amer. Geophys. Union, Washington (1982).
25. C.J. Ebinger, T.D. Bechtel, D.W. Forsyth and C.O. Bowin. Effective elastic plate thickness beneath the East African and Afar plateaus and dynamic compensation of the uplifts, *Jour. Geophys. Res.* **93**, 2883-2901 (1989).
26. P.C. England and G.A. Housemann. The mechanics of the Tibetan Plateau, *Phil. Trans. R. Sos. Lond.* **A 326**, 301- 319 (1988).
27. O.M. Filatov and V.V. Kozlov. On relationships between recent and old tectonic elements and structural forms of the Atlas, *Izvest. Akad. Nauk SSSR, Geol. Ser.* **1**, 78-84 (1974).
28. S.P. Gao, P.M. Davis, M.H. Liu, P.D. Slack, Y.A. Zorin, N.A. Logachev, M.G. Kogan, P.D. Burkholder and R.P. Meyer. Asymmetric upwarp of the asthenosphere beneath the Baikal rift zone, Siberia, *Jour. Geophys. Res.* **96**, 15319-15330 (1994).
29. *Geology of the USSR, Turkmenian SSR*. Nedra, Moscow (1972) (in Russian).
30. A.R. Gilchrist, H. Kooi and C. Beaumont. Post-Gondwana geomorphic evolution of southwestern Africa: Implications for the controls on landscape development from observations and numerical experiments, *Jour. Geophys. Res.* **99**, 12211-12228 (1994).
31. A.R. Gilchrist and M.A. Summerfield. Tectonic models of passive margin evolution and their implications for theories of long-term landscape development. In: *Process models and theoretical geomorphology*. M.J. Kirkby (Ed.). pp. 55-84. John Wiley, New York (1994).
32. A.R. Gilchrist, M.A. Summerfield and H.A.P. Cockbrun. Landscape dissection, isostatic uplift, and the morphologic development of orogenes, *Geology* **22**, 963-966 (1994).
33. A. de Goer de Herve, G. Camus, P. Boivin, A. Gourgaud, G. Kieffer, J. Mergoil and P.-M. Vincent. *Volcanologie de la Chaine des Puys (Massif Central Français) 3eme edition*, I.G.N., Paris (1991).
34. A.M. Goodwin. *Precambrian Geology*. Harcourt Brace Jovanovich, London, San Diego, New York (1991).
35. R.W. Griffiths and I.H. Campbell. Interaction of mantle plume heads with the Earth's surface and onset of small scale convection, *Jour. Geophys. Res.* **96**, 18295-18310 (1991).
36. B.H. Hager and R.W. Clayton. Constraints on the structure of mantle convection using seismic observations, flow models, and the geoid. In: *Mantle Convection*. W.R. Peltier (Ed.). pp 657-766. Gordon and Breach, New York, London, Paris (1989).

37. B.U. Haq, J. Hardenbol and P.R. Vail. Chronology of fluctuating sea levels since the Triassic, *Science* **235**, 1156-1167 (1987).
38. R. Hartley, A.B. Watts and J.D. Fairhead. Isostasy of Africa, *Earth Planet. Sci. Lett.* **137**, 1-18 (1996).
39. C.J.H. Hartnady and T.C. Partridge. Neotectonic uplift in southern Africa: A brief review and geomorphic evidence. In: *Centennial Geocongress Rand Africaans University, Extended Abstracts,* v. **I**. J.M. Barton Jr. and Y.E. Copperthwaite (Eds.). pp. 456-459. Geol. Soc. South Africa (1995).
40. M. He and K. He. The planation surface of the Yunnan Plateau and its neotectonic significance, *Z. Geomorph. N. F., Suppl. - Bd.* **63**, 51-56 (1987).
41. A.R. Jado, H. Hötzl and B. Roscher. Development of sedimentation along the Saudi Arabian Red Sea coast, *Spannung und Spannungsundwandlung in der Lithosphäre, Sonderforschung Bereit* **108 B**, 863-890, Universität Karlsruhe (1989).
42. V.E. Khain. *Regional geotectonics, non-Alpine Europe and Western Asia.* Nedra, Moscow (1977) (in Russian).
43. A.I. Kieselev and A.M. Popov. Asthenospheric diapir beneath the Baikal rift: petrological constraints, *Tectonophysics* **208**, 287-295 (1992).
44. L.C. King. *The morphology of the Earth, 2nd edn.* Oliver and Boyd, Edinburgh (1967).
45. L.C. King. *Wandering continents and spreading sea floors on an expanding Earth.* Wiley, New York (1983).
46. D.L. Kohlstedt, B. Evans and S.J. Mackwell. Strength of the lithosphere: Constraints imposed by laboratory experiments, *Jour. Geophys. Res.*, **100**, 17587-17602 (1995).
47. H. Kooi and C. Beaumont. Escarpment evolution on high-elevation rifted margins: Insights derived from a surface processes model that combines diffusion, advection and reaction, *Jour. Geophys. Res.* **99**, 12191-12209 (1994).
48. H. Kooi and C. Beaumont. Large-scale geomorphology: Classical concepts reconciled and integrated with contemporary ideas via a surface process model, *Jour. Geophys. Res.* **99**, 3361-3386 (1996).
49. G.L. Kosarev, N.V. Petersen, L.P. Vinnik and S.W. Roecker. Receiver functions for the Tien Shan analogue broad band network: Contrasts in the evolution of structures across the Talassa-Fergana fault, *Jour. Geophys. Res.* **98**, 4437-4448, (1993).
50. V.N. Krestnikov. *The history of evolution of oscillating crustal movements in the Pamirs and adjacent regions in Asia.* Akad. Sci. USSR Press, Moscow (1962) (in Russian).
51. Li Jijun et al. *Uplift of the Qinghai-Xizang (Tibet) Plateau and global change.* Lanzhou Univ. Press (1995).
52. B.A. Litvinovsky, E.V. Artyushkov and A.N. Zanvilevich. On the nature of magmatism of the Mongolian - Transbaikalian Belt, *Geologiya i Geofisika* **2**, 32-40 (1989) (in Russian).
53. N.A. Logachev and S.I. Sherman (Eds.). *Regional Neotectonics of Siberia.* Nauka, Novosibirsk (1983) (in Russian).
54. N.V. Lukina. Relief of mountain belts of southern USSR as an indicator of the Neotectonic vertical and horizontal movements. In: *Geodynamics of Intraplate Mountain Regions.* N.A. Logachev (Ed.). pp. 84-90, Nauka, Novosibirsk (1990) (in Russian).
55. V.I. Makarov. *Recent orogenes, their structure and geodynamics.* Geological Institute, Moscow (1990) (in Russian).
56. V.I. Makarov. Neotectonics and geodynamics of mountain systems of Central Asia, *Quaternary Intern.* **25**, 19-23 (1995).
57. E.E. Milanovsky. *Neotectonics of the Caucasus.* Nedra, Moscow (1968) (in Russian).
58. E.E. Milanovsky. Geomorphology. In: *East-African Rift System*, v. **2**. V.V. Beloussov, A.V. Goryachev, and E.E. Milanovsky (Eds.). pp. 56-130. Nauka, Moscow (1974) (in Russian).
59. V.S. Mileev, S.B. Rosanov and E.Yu. Baraboshkin. The position of the Upper Jurassic deposits in the structure of the Mountain Crimea, *MOIP Bull., ser. Geology* **70** (**1**), 22-31 (1995) (in Russian).
60. A. Miyashiro, K. Aki and A.M.C. Sengör. *Orogeny.* Wiley, Chichester (1982).
61. L.I. Mokeeva, L.P. Vinnik and S.W. Roecker. Shear wave splitting and small scale convection in the continental upper mantle, *Nature* **358**, 144-146 (1992).
62. P. Molnar and P.C. England. Late Cenozoic uplift of mountain ranges and global climate change: chicken or egg, *Nature* **346**, 29-34 (1990).
63. P. Molnar and P. Tapponnier. Cenozoic tectonics of Asia: effects of continental collision, *Science* **189**, 419-426 (1975).
64. M. Monnereau, M. Rabinovicz and E. Arquis. Mechanical erosion and reheating of the lithosphere: a numerical model for hotspot swells, *Jour. Geophys. Res.* **98**, 809-824 (1993).
65. J.F. Morgan, W.J. Morgan and Yu. Zhang. Observational hints for a plume-fed, suboceanic mantle, suboceanic asthenosphere and its role in mantle convection, *Jour. Geophys. Res.* **100**, 12753-12767 (1995).

66. N.-A. Mörner. Neotectonics, the new global tectonic regime during the last 3 Ma and the initiation of ice ages, *An. Acad. bras. CL* **65** (supl. **2**), 295-301 (1993).
67. *A New Earth Gravity Field Model in Support of ERS-1 and SPOT-2: GRIM 4-SI/CI*. Final Report of the German Space Agency (DARA) and the French Space Agency (CNES) (1991).
68. N.I. Nikolajew. *Neotectonics and its Expression in the Structure and Relief of the USSR Territory*, Gostekhizdat, Moscow, 1962 (in Russian).
69. N.I. Nikolajew, Yu.Ya. Kuznetsov and A.A. Neumark (Eds.). *Neotectonic map of the world, scale 1:15 000 000*. Ministry of Geology of the USSR, Ministry of Higher and Secondary Special Education of the USSR (1981).
70. N.I. Nikolajew, Yu.Ya. Kuznetsov and A.A. Neumark (Eds.). *Neotectonics of continental, transitional and oceanic regions of the world (explanatory note to the Neotectonic map of the world)*. VNIIZarubezhgeologiya (1984) (in Russian).
71. N.I. Nikolajew and A.A. Neumark (Eds.). *Neotectonic map of the USSR and the adjoining areas, scale 1:5 000 000*. Ministry of Geology of the USSR, Ministry of Higher and Secondary Special Education of the USSR, Moscow (1977).
72. A.A. Nyblade and H.N. Pollack. A global analysis of heat flow from Precambrian terrains; implications for the thermal structure of Archean and Proterozoic lithosphere, *Jour. Geophys. Res.* **98**, 12207-12218 (1993).
73. C.D. Ollier. *Ancient landforms*. Belhaven, London (1991).
74. T.C. Partridge and R.R. Maud. Geomorphic evolution of southern Africa since the Mesozoic, *S. Afr. Jour. Geol.* **90**, 179-208 (1987).
75. J.A. Pearce and Mei Houjin. Volcanic rocks of the 1985 Tibet geotraverse: Lhasa to Golmud, *Phil. Trans. R. Soc. Lond.* **A 327**, 169-201 (1988).
76. *Plateau uplift*. K. Fuchs et al. (Eds.), Springer, Berlin, Heidelberg (1983).
77. J. Polet and D.L. Anderson. Depth extent of cratons as inferred from tomographic studies, *Geology* **23**, 205-208 (1995).
78. K. Priestley, J. Cipar, A. Egorkin and N. Pavlenkova. Upper-mantle velocity structure beneath the Siberian platform, *Geophys. Jour. Int.* **118**, 369-378 (1994).
79. M.E. Raymo and W.F. Ruddiman. Tectonic forcing of late Cenozoic climate, *Nature* **359**, 117-122, 1992.
80. N.M. Ribe and U.R. Christensen. Three-dimensional modeling of plume-lithosphere interaction, *Jour. Geophys. Res.* **99**, 669-682 (1994).
81. S.W. Roecker, T.M. Sabitova, L.P. Vinnik, Y.A. Burmakov, M.I. Golvanov, R. Mamatkanova, and L. Minurova. Three-dimensional elastic wave velocity structure of the western and central Tien Shan, *Jour. Geophys. Res.* **98**, 15779-15795 (1993).
82. W.F. Ruddiman, W.L. Prell and M.E. Raymo. Late Cenozoic uplift in southern Asia and American west: Rationale for general circulation modeling experiments, *Jour. Geophys. Res.* **94**, 18379-18391 (1989).
83. E.P. Saggerson and B.H. Baker. Post-Jurassic erosion-surfaces in eastern Kenya and their deformation in relation to rift structure, *Quat. Jour. Geol. Soc. Lond.* **121**, 51-72 (1965).
84. I.G. Scherba. *Stages and phases of the Cenozoic evolution of the Alpine region*. Nauka, Moscow (1993) (in Russian).
85. S.S. Schulz. *Analysis of the neotectonics and the relief of the Tien Shan*. OGIZ, Moscow (1948) (in Russian).
86. D. Seber, M. Barazangi, A. Ben Tadili, M. Ramdani, A. Ibenbrahim and D. Ben Sari. Three-dimensional upper mantle structure beneath the intraplate Atlas and interplate Rif mountains of Morocco, *Jour. Geophys. Res.* **101**, 3125-3138 (1996).
87. N.K. Sleep. Lithospheric thinning by midplate mantle plumes and the thermal history of hot plume material ponded at sublithospheric depth, *Jour. Geophys. Res.* **99**, 9327-9343 (1994).
88. Ya.B. Smirnov. *Geothermic map of North Eurasia and methods of analysis of the thermal structure of the lithosphere (explanatory text)*. Geol. Inst. Acad. Sci. USSR, Moscow (1986) (in Russian).
89. B.S. Sokolov. On the stratigraphy of the Tertiary and Early Quaternary deposits of the Naryn depression (Tien Shan), *Dokl. Akad. Nauk SSSR* **LXVII**, 681-684 (1949) (in Russian).
90. M.S. Steckler. Uplift and extension of the Gulf of Suez: indications of induced mantle convection, *Nature* **317**, 135-139 (1985).
91. M.P. Strecker, P. Cerveny, A.L. Bloom and D. Malizia. Late Cenozoic tectonism and landscape development in the foreland of the Andes: Northern Sierras Pampeanas (26°-28°S), Argentina, *Tectonics* **8**, 517-534 (1989).
92. J. Strehlau and R. Meissner. Estimation of crustal viscosities and shear stresses from an exploration of experimental steady state flow data. In: *Composition, structure and dynamics of the lithosphere-asthenosphere system*. K. Fuchs and C. Froidevaux (Eds.). pp. 67-87. Amer. Geophys. Un., Washington (1987).

93. W. Su, R.L. Woodward and A.M. Dziewonsky. Degree 12 model of shear velocity heterogeneity in the mantle, *Jour. Geophys. Res.* **99**, 6945-6980 (1994).
94. M.A. Summerfield. *Global geomorphology: an introduction to the study of landforms.* Burnt Mills, Longman (1991).
95. V.G. Trifonov (Ed. in chief). *Map of active faults of Northern Eurasia. Scale 1: 5.000.000.* Russian Comm. on the Lithosphere, Moscow (1995).
96. L.P. Vinnik, R.W.E. Green and L.O. Nicolaysen. Seismic constraints on dynamics of the mantle of the Kaapvaal craton, *Phys. Earth Planet. Inter.* **95**, 139-151 (1996).
97. L.P. Vinnik, R.W.E. Green, L.O. Nicolaysen, G.L. Kosarev and N.V. Petersen. Deep seismic structure of the Kaapvaal craton, *Tectonophysics*, in press.
98. J.K. Weissel and G.D. Karner. Flexural uplift of rift flanks due to mechanical unloading of the lithosphere during extension, *Jour. Geophys. Res.* **94**, 13919-13950 (1989).
99. M. Wilson and R. Guiraud. Magmatism and rifting in Western and Central Africa, from Late Jurassic to Recent times, *Tectonophysics* **213**, 203-225 (1992).
100. D.A. Yuen and L. Fleitout. Thinning of the lithosphere by small-scale convective destabilization, *Nature* **313**, 125-188 (1992).
101. P.K. Zeitler. Cooling history of the NW Himalayas, Pakistan, *Tectonics* **4**, 127-151 (1985).
102. Zhang Benxing. The influence of Himalayan uplift on the development of Quaternary glaciers, *Z. Geomorph. N. F., Suppl. - Bd.* **76**, 89-115 (1989).
103. Yu.A. Zorin and N.A. Florensov. On geodynamics of Cenozoic uplifts in Central Asia, *Tectonophysics* **61**, 271-283 (1979).

Proc. 30th Int'l Geol. Congr., Vol.7, pp. 121-138
Xu Zhiqin, Ren Yufeng and Qiu Xiaoping (Eds.)

Neoproterozoic and Late Paleozoic Taimyr Orogenic and Ophiolitic Belts, North Asia: A Review and Models For Their Formation

VALERY A. VERNIKOVSKY
United Institute of Geology, Geophysics and Mineralogy, Siberian Branch, Russian Academy of Sciences

Abstract

This article contains two parts: 1) tectonic setting, petrology, geochemistry and geochronology of Taimyr ophilolitic, island arc and collisional complexes, and 2) an application of the evolution of Taimyr Neoproterozoic and Late Paleozoic mobile belts to the Arctic tectonics.

Tectonic studies and petrological analysis suggest that the Neoproterozoic Central Taimyr accretionary belt is located between the Kara and Siberian continents. This belt includes the Mamont-Shrenk and Faddey gneissic terranes with 850 Ma granites (model Sm-Nd age is 1800-1900 Ma), and island arc units, and ophiolites with 740 Ma plagiogranites (model Sm-Nd age is 785-850 Ma).

The collision of the accretionary block with the Siberian continent might have occurred 600 Ma which is the age of metamorphism of garnet amphibolites from the Stanovoy belt (Sm-Nd, Ar-Ar, Rb-Sr and K-Ar method). This event resulted in thrusting, metamorphism, granitization, and uplift hight-pressure complexes.

The next significant tectonic event was Late Carboniferous-Permian collision of the Kara and Siberian continents and formation of the Kara orogenic belt with enormous volumes of zonally metamorphosed rocks, syn-collisional (300 Ma) and post-collisional (264 Ma) granites, and transformations of the Siberian continental margin.

The models for the formation of ophiolitic belt and Kara collisional orogenic belt, and proposed paleogeodynamic reconstruction are discussed.

key words: accretionary belt, ophiolite, metamorphism, collision

INTRODUCTION

The Taimyr fold area has a key position among the main structures of the Arctics. The South-Taimyr foldbelt (mainly Paleozoic-Early Mesozoic) is situated between the Ural-Novozemelsky foldbelt and the Verkhoyansk-Chukotka fold area. The northern part of Taimyr is considered by some authors as a block which together with structures of Novosibirsk Island, Chukotka Peninsula, Sjuard Peninsula, Canadian Archipelago, and North Greenland made up the Arctida paleocontinent [1-3].

In addition, study of the main structures and formation of the Taimyr fold area is significant to the explanation of geological evolution of the Siberian Craton which outlines the Taimyr structures from the north.

Up to now a large number of concepts to the nature of Taimyr have appeared. Some involve the schemes for the ancient geosyncline evolution (from the Archean to Paleozoic), others are based on the Plate tectonic theory. The main investigations that significantly formed concepts on the geology and evolution of Taimyr are as follows: [2, 4-8].

Recently, three structural units occurring parallel to the general trend of the orogenic zone have been distinguished in the Taimyr fold area. Each unit has its own characteristic of the composition and individual Precambrian-Paleozoic history [2,9,10]. One of them includes Neoproterozoic ophiolitic belts which are interesting in the contexts of their correlation with similar complexes of the south-west framing of the Siberian Craton and the Arctics. These ophiolites possibly record the opening of the Paleopacific Ocean and the breakup of Neoproterozoic Rodinia supercontinent.

Formation of the Kara orogenic belt is important for the understanding of the Arctics development in the Paleozoic-Mesozoic. This event was coeval with the formation of Pangea, in the Late Paleozoic.

GEOLOGIC AND TECTONIC SETTING

The Taimyr fold area is traced up to 1000 km, at a width more than 300 km, along the Kara sea shore. Its southern boundary is hidden under Meso-Cenozoic sediments of the Yenisei-Khatanga trough, and its northern part is overlapped by waters of the Kara Sea. Three zones are distinguished within the Taimyr fold area: South-, Central-, and North-Taimyr, whose structural and compositional features suggest various geodynamic conditions of their formation (Fig. 1). They are separated by the largest thrust structures of the region - Pyasino-Faddey, Main Taimyr, and Diabasovy.

The South-Taimyr Paleozoic-Mesozoic fold zone is composed of unmetamorphosed Ordovician-Permian sedimentary deposits of the deep-sea trough and the Late-Permian - Early Triassic volcanogenic-sedimentary deposits hosting flood basalt sheets and dolerite dikes and sills of trappean formation. All deposits of the South-Taimyr zone are dislocated with significant decrease in the intensity of both disjunctive and plicative dislocations in a southward direction.

The Central-Taimyr zone is, probably, the most complex in structure as compared to others two zones. Its structure includes the Pre-Neoproterozoic (?) and Neoproterozoic sedimentary, volcanogenic, and intrusive formations of various age which underwent metamorphic and hydrothermal-metasomatic alterations of various facies and types. These metamorphic formations are combined into an accretion belt and overlapped by Vendian - Early Carboniferous cover.

The Pyasino-Faddey overthrust is, as mentioned above, a boundary structure between the Central- and South-Taimyr zones. This is the most extended sutural zone of Taimyr (about 1000 km in length and 30-40 km in width). It consists of smaller echelon-like located heteroamplitudinal overthrusts [7] or is a complex system of alternation of plates, flakes, and tectonic wedges, along which the essentially terrigenous relatively deep-sea

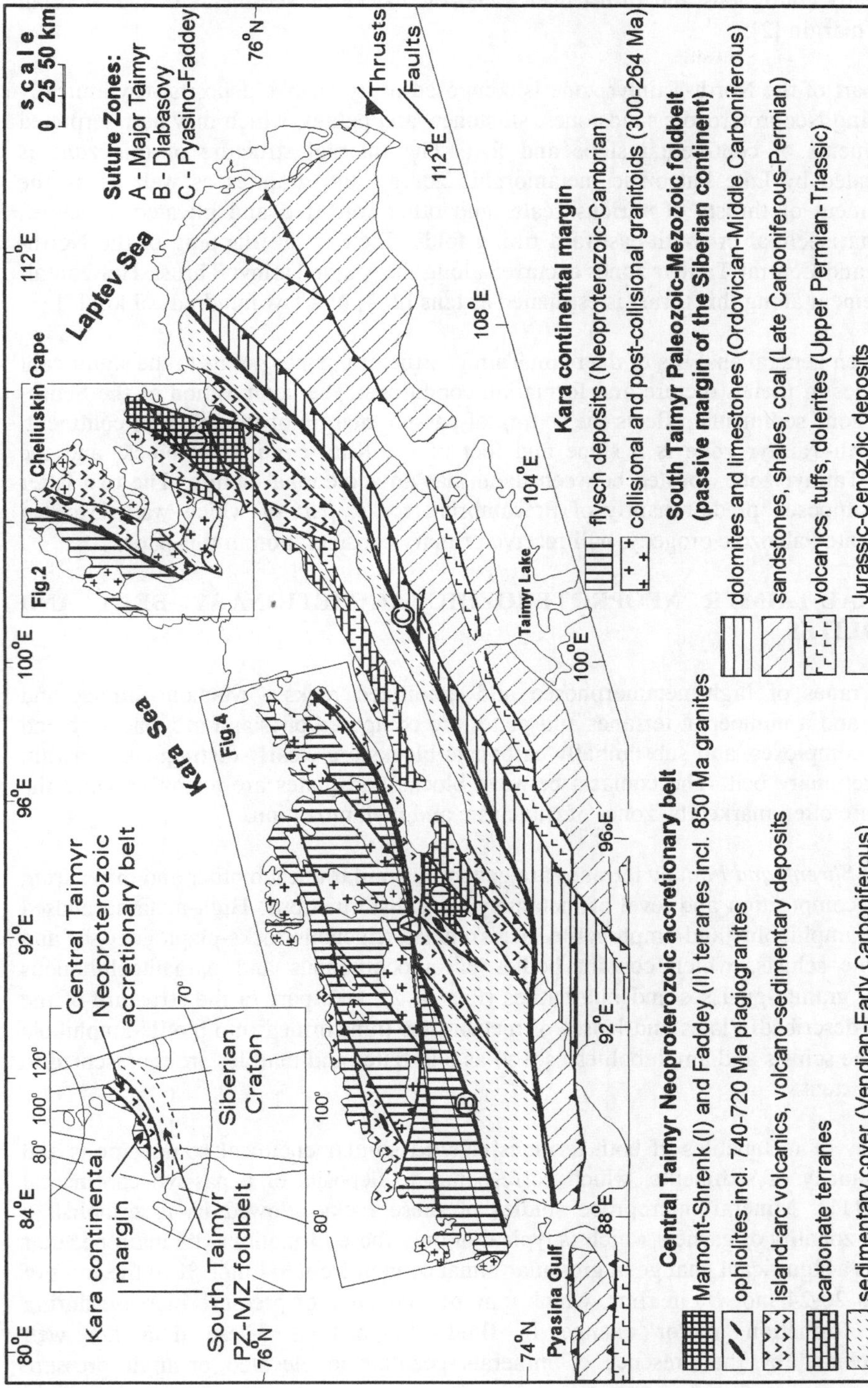

Fig. 1. Tectonic map of the Taimyr folded area

sediments of sedimentary cover of the accretion block (Central-Taimyr zone) are tectonically contiguous and thrust over Paleozoic carbonate sediments of the Siberian passive margin [2].

Major part of the North-Taimyr zone is composed of flysch-like deposits -rhythmically alternating Neoproterozoic sandstones, siltstones, and pelites, which may be interpreted as sediments of continental slope and foot. The internal structure of this zone is complicated by Late Paleozoic metamorphic zoning and granites as well as by the development of thrusts of various scale, and other fractured and plicated structures including isoclinal or south-eastward tilting folds. Tectonic overlapping of the North-Taimyr and Central-Taimyr zones occurred along the Major Taimyr Thrust. Horizontal displacement along this thrust is estimated as tens [2, 4] or a few hundreds of km [7].

Thus, even general analysis of the main Taimyr structural zones points to the significant differences in their structure and formation conditions. The composition of the South-Taimyr zone sediments reflects the setting of passive margin of the Siberian continent. The North-Taimyr zone is a slope and foot of the Kara continental block, and the Central-Taimyr zone, located between them, has an accretional nature. The two latter zones composed predominantly of Precambrian rocks, part of which was reworked during Late Paleozoic orogeny, will receive primary consideration in this paper.

CENTRAL-TAIMYR NEOPROTEROZOIC ACCRETIONARY BELT AND OPHIOLITES

Two terranes of high-metamorphosed and granitized rocks - Mamont-Shrenk and Faddey, and a number of terranes and plates are composed of island-arc, back-arc and oceanic complexes, and substantially carbonate blocks are clearly distinguished within the accretionary belt. The contacts between blocks and plates are tectonic, while the thrusts are often marked by zones of cataclasm and mylonitization.

Mamont-Shrenk and Faddey terranes are generally similar to each other and differ from other in composition and level of metamorphic transformations. High-metamorphosed (epidote-amphibolite and amphibolite facies) metaterrigenous rocks-plagiogneisses and crystalline schists, which contain bodies of autochthonous and paraautochthonous granites, granite-gneisses and migmatites (Fig. 1, 2), take part in the structure of the terranes described. Plates and lenses of metabasites transformed into biotite-amphibole crystalline schists and amphibolites as well as quartzites and marbles are represented to a lesser extent.

The study of metapelites of both terranes showed similar chemical compositions and high maturity of sediments, which corresponds to deposits of a passive continental margin [11]. Mineral microprobe studies of these rocks allowed us to establish a complex zonality of garnets which is typical of only these formations. Its main peculiar feature is a significant change in grossular minal content from 3-4 mol. % in the cores of grains to 20-24 mol. % in rims, which may be indicative of pressure increase during mineral formation and/or change in fluid composition. These data are well complemented by the presence of minerals peculiar to elevated or high pressure

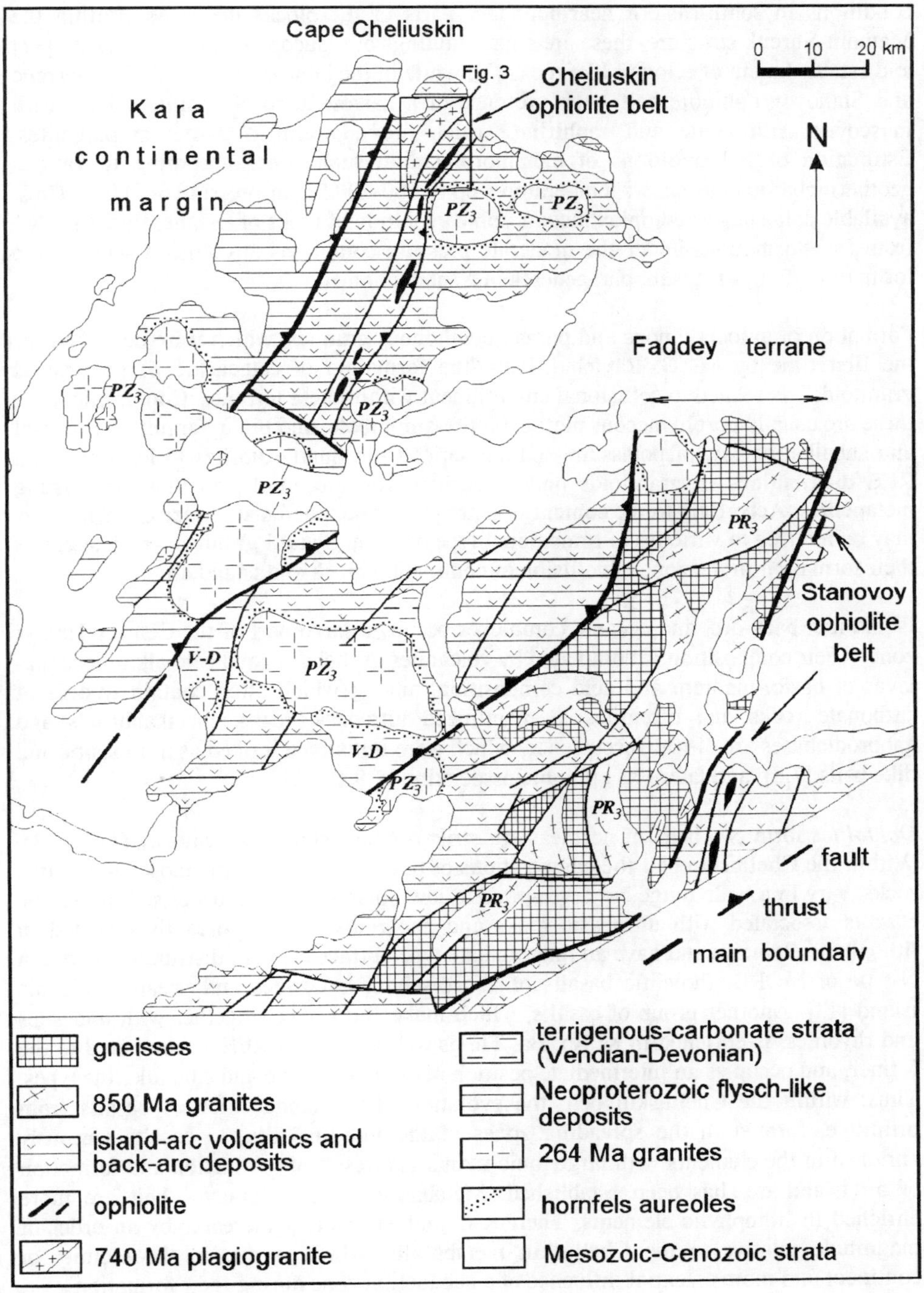

Fig. 2. Geological map of North-Eastern Taimyr

conditions in south-eastern near-boundary parts of the blocks described. Within the Mamont-Shrenk structure, these are single findings of glaucophane and omphacite [12] and establishment of eclogite-like rocks [7], while in the joint zone of the Faddey terrane and Stanovoy ophiolite belt - the presence of a narrow zone of kyanite schists with muscovite, sillimanite and saphirine, which are adjacent to garnet amphibolites. Estimation of PT-conditions of metamorphism in such zones according to various geothermobarometers shows temperatures up to 600-700 °C, at pressure 6-9 kbar. Thus, available data suggest complex metamorphic evolution of rocks of terranes under study - from transformations under the moderate pressure conditions at collision stage to the formation of high-pressure parageneses in a subduction zone.

Formation of autochthonous and paraautochthonous granites, which were described for the first time by M. G. Ravich [5] as "the formation of cataclastic porphyraceous granitoids", is related to collisional environment within these terranes. Compositionally, these are usually porphyraceous biotite, biotite-amphibole, and binary granites of normal and subalkaline raw of potassium-sodium series from quartz diorites to leucogranites. REE distributions in granitoids under consideration are similar to those in hosting metapelites. According to the content of petrogenic components and trace elements they may be referred to varieties of transitional type between S and I granites which suggests their formation in the zone of collision of continental block and island arc.

Island-arc, back-arc, and oceanic complexes occupy a large part of the Central-Taimyr zone. Their composition is dominated by volcanites of tholeiitic and calc-alkaline seria - lavas of basic, medium and acid composition, among which often contain wedges of carbonate rocks and turbidites. Serpentinized ultrabasic rocks, metagabbroids and gabbrodiabases are of wide occurrence, which were initially described as ultrabasitic and dike belts [7,8], and later - as ophiolite association [2,9,10,13].

Ophiolites form two belts of northeastern strike - Cheliuskin and Stanovoy (Fig. 2, 3). Within the Cheliuskin belt the compositions of basalts, which are the most informative rocks, vary in a wide range but correspond to low-potassium tholeiites of sodium series. Basalts associated with ultrabasic rocks and gabbroids, are significantly depleted in lithophylic elements and have an intermediate type of trace element distribution between N-type of MORB, tholeiitic basalts of above-subduction zones and oceanic volcanic island [13]. Another group of basalts, which makes up a sheet together with andesites and rhyolites, is enriched in K, Rb, Ba, Th, as well as in light REE as compared to the former, and occupies an intermediate position between tholeiitic and calc-alkaline types. Thus, within the Cheliuskin belt, the evolution of the composition of basalts from primitive, formed in the spreading zones of the marginal sea, to basalts variously enriched in the elements with large ionic radius as a result of subduction and formation of an island-arc, has been established. Metabasalts of the Stanovoy belt are more enriched in lithophylic elements. Their K_2O and Rb content increases by an order of magnitude relative to the Cheliuskin metabasalts, which is due to the vicinity of continent and more essential influence of a subduction zone during their formation.

Plagiogranites of tonalite-trondhjemite series, which compositionally similar to plagiogranites of typical ophiolite complexes, described as oceanic plagiogranites [14], continue the low-potassium series of basic rocks of the Cheliuskin belt. They make up plate-like bodies with visible thickness up to 1 km among metabasalts, andesito-basalts, and metagabbroids (Fig. 3). Plagiogranites are characterized by low concentrations of K_2O, Rb, Nb, U, Th, elevated - Y and Zr, and REE distribution in them is similar to that of average basalt of the Mid-Atlantic Oceanic Ridge, as well as for gabbroids, basalts, and andesito-basalts of the ophiolite complex in question [15].

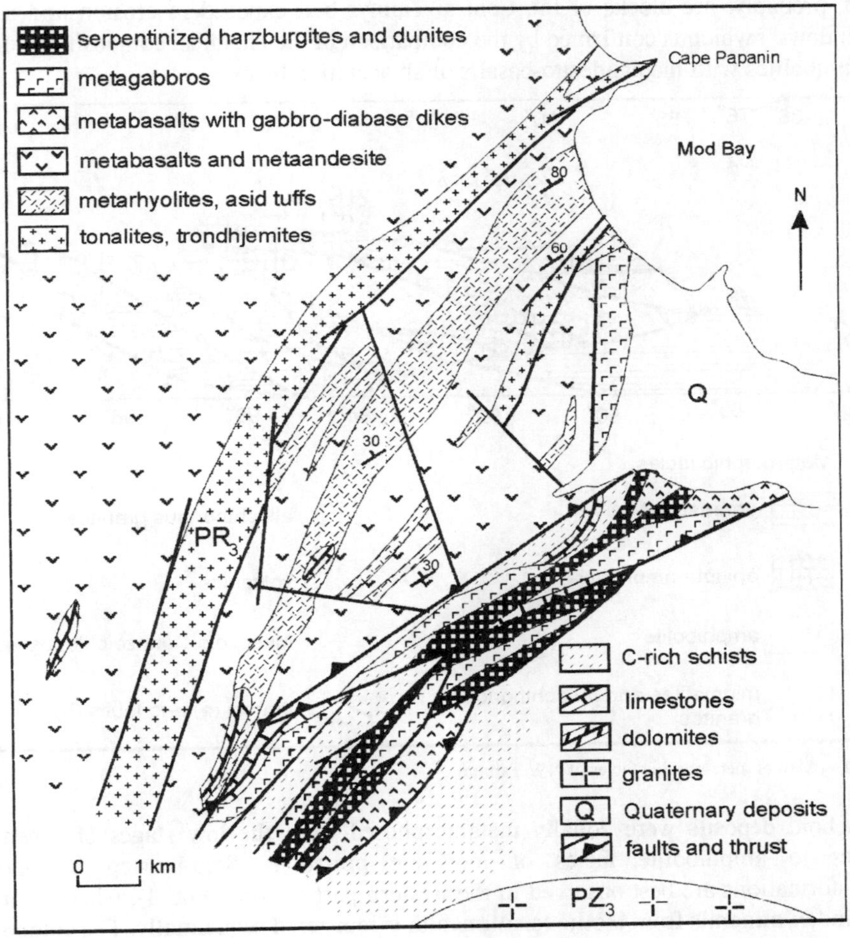

Fig 3. Geological map of the NE Cheliuskin ophiolitic belt

Rocks of the oceanic, island-arc and back-arc complexes were regionally metamorphosed from greenschist facies within the Cheliuskin belt to amphibolite facies of elevated pressures in the Stanovoy belt. In the latter case, garnet amphibolites are developed after basalts. This fact, to my mind, together with other factors mentioned, suggests transformations in the paleozone of subduction.

KARA CONTINENTAL MARGIN

Metamorphosed rhythmically alternating sandstones, siltstones, and pelites of continental slope and foot occupy a considerable part of the North-Taimyr zone. Granitoids of autochthonous, paraautochthonous and allochthonous types are localized among flyschoid sediments. Blocks of amphibolites and biotite-amphibole crystalline schists are present in a lesser volume. They have tectonic contacts with flyshoid deposits and, probably, are blocks of the Central-Taimyr belt exposed in erosion and tectonic "windows", which is confirmed by the comparison of the chemical compositions of these amphibolites with metaandesito-basalts of an accretion block.

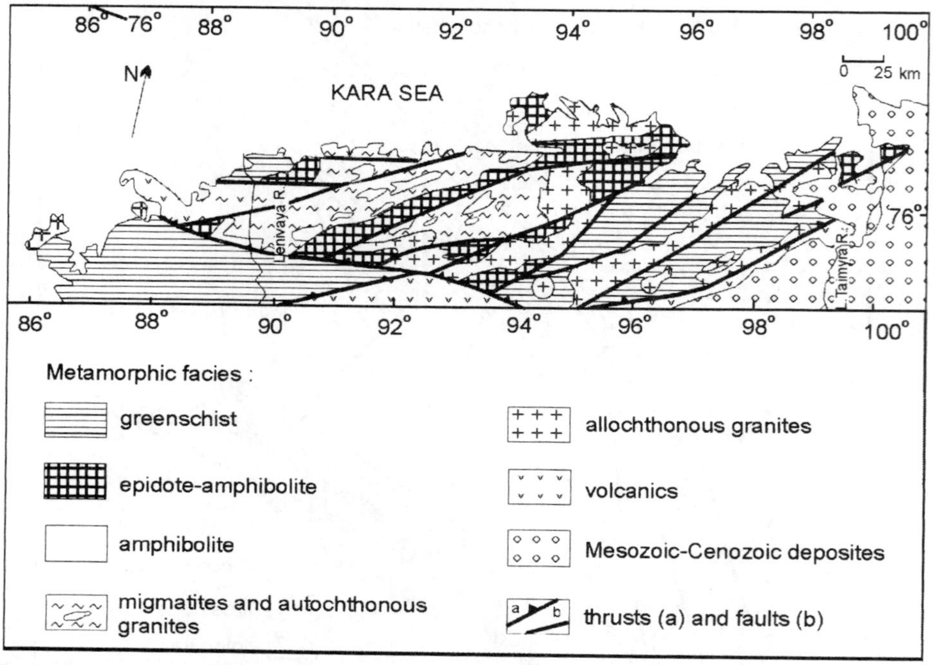

Fig. 4 Map of metamorphic facies of NW Taimyr

Flyschoid deposits were zonally metamorphosed from the low stages of greenschist facies to amphibolite facies of moderate pressures. Step-by-step metamorphic transformations are best observed in the western part of zone (Fig. 4), where change of the index-minerals from biotite to sillimanite is registered sequentially. The composition of metamorphic minerals is changed successively as well. In particular, a decrease in the content of spessartine minal in garnet from low-temperature to high-temperature zones from 15-16 to 3-4 mol.% is accompanied by a rise of pyrope component share from 8-9 to 19-21 mol.%. Estimation of PT-parameters of metamorphism by different geothermobarometers showed that temperatures from garnet to sillimanite zone vary from 460 to 650 °C, and pressures - from 3 to 6.5 kbar [11].

Metamorphic transformations of amphibolite facies are often accompanied by migmatization and granitization with formation of elongated bodies of autochtonous biotite and binary porphyroblastic gneiss-granites, subalkaline granites and quartz syenites. Paraautochthonous and allochthonous granitoids occur among metamorphites of epidote-amphibolite and greenschist facies. Autochthonous varieties occur only in the western part of the region (Fig. 4), while clearly discordant massifs are typical of the Cheliuskin Peninsula (Fig. 2). The latter break out both regionally metamorphosed formations (flyschoid ones of the Kara continent and volcanogenic-sedimentary of an island arc), and unmetamorphosed Paleozoic deposits of accretion block sheet with formation of haloes of contact hornfels. The substance composition of the allochthonous massifs is more various than that of the above mentioned autochthonous varieties. They contain varieties of porphyraceous, biotite-hornblende granites, granodiorites, granosyenites, quartz syenites, diorites, and monzonites.

The petrochemical characteristics along with the peculiar features of trace element composition allow granitoids of the North-Taimyr zone to be treated as the crustal granites of calc-alkaline series of transitional type between S and I granite [16]. Comparison of trace-element composition of autochthonous and allochthonous varieties suggests a decrease in Rb content in latter with an increase in Sr and Ba content, which is specific of postcollisional granites [17]. Specific features of various granite types are also established when analyzing REE distribution in them. The geochemical properties of autochthonous granites were to a greater degree inherited from rocks of substratum. The La/Yb ratio drastically increases in allochthonous varieties that suggests clear REE fractionation, i.e. predominant remove of heavy lanthanoids.

Morphology of the massifs, relationships between granites and host rocks, and their chemical composition indicate that some of them (autochthonous and paraautochthonous varieties) were generated under syncollision conditions, while others (allochthonous massifs of the North-Eastern Taimyr) - under postcollisional or orogenic conditions.

ISOTOPIC GEOCHEMISTRY AND EVOLUTION OF CONTINENTAL CRUST

The main discussions on the problems of tectonic evolution of the Taimyr fold area were mainly governed by nearly total absence of precise isotopic-geochronological data for magmatic and metamorphic complexes in this region. Published K-Ar datings in general correspond to the interval of 280-225 Ma and reflect thermal processes of Late Paleozoic orogeny and Early Mesozoic stage of trappean magmatism [18]. The available scarce isotopic data on crystalline complexes, obtained by Pb-Pb and U-Pb-Th methods, were published without any calculated data which did not suggest their reliability [7,19]. Interpretation of U-Pb data on zircons from granites of Mamont-Shrenk terrane [20] turned out to be ambiguous. In this case, lower intersection of discordia with concordia corresponds to 612±43 Ma, while upper - to 1869±56 Ma, but all figurative dots are grouped close to the lower intersection. Hence, it is possible that the Early Proterozoic age corresponds not to granite crystallization, but to the age of ancient inherited component of radiogenic Pb. Moreover, the same authors determined U-Pb age of sphene to be 817±30 Ma in the granites studied, while K-Ar method for muscovite showed values 803±23 and 813±26 Ma. Similar results were obtained by Rb-Sr isochron method in determining the age of metamorphism of gneiss, hosting these granites - 763±12 Ma [21].

Within the last years we together with the collegues of the Institute of Precambrian Geology and Geochronology (Sankt-Petersburg) and the United Institute of Geology, Geophysics and Mineralogy (Novosibirsk) carried out complex isotopic-geochronological studies of key magmatic and metamorphic complexes of Taimyr using U-Pb, Sm-Nd, Rb-Sr, Ar-Ar and K-Ar methods (Table 1). Sm-Nd isotope-geochemical studies showed that formation of both Neoproterozoic and Late Paleozoic granites of Taimyr occurred with the participation of Precambrian crust material (Fig.5). The most ancient Sm-Nd model age of the continental crust was established in gneiss-granite Faddey terrane - 1796-1902 Ma. Granites of the Zhdanov Massif were originated at the cost of reworking of Early Proterozoic crust substratum in Neoproterozoic (850 Ma) or either as the result of microcontinent - continent or island arc - microcontinent collision. The age datings for plagiogranites of the Cheliuskin ophiolite belt are similar to the boundary of this Neoproterozoic collision, which suggests interaction of the island arc and microcontinent. Study of various isotopic systems in garnet amphibolites of the Stanovoy ophiolite belt allowed us to establish the age interval of their metamorphism which probably corresponds to the boundary of obduction of ophiolites and the whole accretion block to the margin of the Siberian craton - 626-575 Ma.

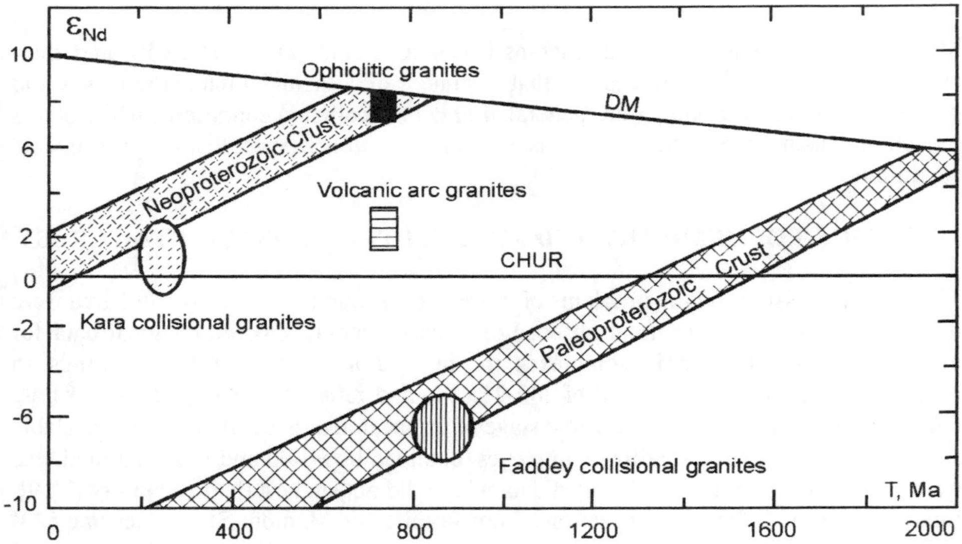

Fig 5. ε_{Nd}– T diagram for granites of Taimyr
DM - depleted mantle; CHUR - mantle with chondritic REE correlation. Evolutionary trends for isotopic Nd composition in the models for continental crust with the age ~800 and ~1900 Ma and average $^{147}Sm/^{144}Nd = 0.12$.

Table 1. Isotopic-geochronological data for granitoids and metamorphic rocks of the Taimyr folded area

Dated object	*Method*	*Age (Ma)*	*Interpretation*	*References*
		Faddey terrane, Zhdanov Granite		
wr	$Sm\text{-}Nd_{TDM}$	1796-1902	Age of crust	[22]
zircon	U-Pb	846±11	Age of crystallization	
		Cheliuskin ophiolites belt, Kunar Plagiogranite		
wr	$Sm\text{-}Nd_{TDM}$	785-850	Age of crust	
zircon	U-Pb	740±38	Age of crystallization	[13, 15]
wr (n=5)	Rb-Sr, isochron	727±83	Age of crystallization	
		Stanovoy ophiolite belt, Garnet amphibolite		
gar, bt, pl, amf, wr	Sm-Nd, isochron	573±78	Age of metamorphism	
gar, bt, pl, amf, wr	Rb-Sr, isochron	606±44	Age of metamorphism	
bt	Ar-Ar	624±16	Age of metamorphism	[23]
bt	K-Ar	596±6	Age of metamorphism	
amf	K-Ar	626±6	Age of metamorphism	
		Kara continental margin, syn-collisional complex, Granite		
wr	$Sm\text{-}Nd_{TDM}$	1170-1080	Age of crust	
zircon, monazite	U-Pb	306±2	Age of crystallization	[16]
bt-flsp-wr	Rb-Sr, isochron	277+4	Cooling <500°C during uplift	
stv-bt-flsp-wr	Rb-Sr, isochron	269±8	Cooling <500°C during uplift	
		Post-collisional complex, Granodiorite		
wr (n=3)	$Sm\text{-}Nd_{TDM}$	850-1075	Age of crust	
zircon	U-Pb	264±8	Age of crystallization	[16, 24]
wr (n=5)	Rb-Sr, isochron	258±28	Cooling <500°C during uplift	
bt	Ar-Ar	252.5±1	Cooling <300°C	

Abbreviations: **amf - amphibolite; bt - biotite; flsp - feldspar; gar - garnet; pl - plagioclase; stv - staurolite; wr - whole rock.**

Quite different datings were obtained during isotopic studies of Late Paleozoic syn- and postcollision granitoids. Syncollision (autochthonous) granites originated at the boundary of 300 Ma at the expense of continental crust of Grenwillian age - T_{Nd}(DM)=1170 Ma, T_{Nd}(DM-2)=1080 Ma. It is interesting that the age of crustal source for postcollisional granitoids is somewhat less than for syncollisional ones and approaches the model estimations of the crustal source for plagiogranites of the Cheliuskin ophiolite belt. These facts as well as mixed composition of postcollisional granitoids in question, corresponding to the transitional type between S and I granites, suggest that they are products of mixing of magmas of acid and basic compositions, which were generated during melting of crustal substrats of Grenwillian (Karacontinental margin) and Neoproterozoic (ensimatic island arc and ophiolites) age.

MODELS FOR FORMATION OF THE OPHIOLITIC AND OROGENIC BELTS

The presence of Neoproterozic ophiolites within the Central Taimyr accretionary belt suggest that in the Late Precambrian time the Siberian and Kara paleocontinents were divided by the oceanic-type crust. At present it is very difficult to determine the size of this oceanic space due to the absence of more or less reliable paleomagnetic data for the North Taimyr. Nevertheless, this can be judged from geodynamic conditions of formation of ophiolites and island arc volcanites. Their structural and compositional features suggest that Cheliuskin and Stanovoy ophiolites are related with subduction and spreading zones of the marginal sea [13].

The model of evolution of ensimatic marginal basins [25] can be used for reconstruction of the Cheliuskin belt. According to this scheme, immediately after beginning of subduction of one oceanic plate under another, hydration of harzburgite mantle adjacent to submerged lithospheric plate takes place causing its partial melting. Melting of the mantle source results in pre-arc spreading and formation of ophiolites including rocks of gabbro-diorite-trondhjemite series. Then comes submarine arc volcanism with formation of, as in our case, series of primitive volcanites from tholeiitic basalts to tholeiitic rhyolite (Fig.6, A, B), compositionally similar to ensimatic Tonga-Kermadek segment [26, 27]. At the next stages of formation of the Cheliuskin ophiolitic belt, closing of spreading centre, scaling of plates and island-arc crust, accumulation and accretion to the blocks of continental masses take place (Fig.6, C, D). At these stages of sheet formation, zones of cataclasm, mylonitization and serpentinite melange are generated. The degree of metamorphism of exposed rocks does not exceed high-temperature zone of greenschist facies. The dating of zircons from plagiogranites (740±38 Ma) and their Rb-Sr isochron age (727±83 Ma) allow us to consider the level of 727-740 Ma to be the upper age boundary of formation of Cheliuskin ophiolites, and Sm-Nd model ages (850-785 Ma), probably, characterize their lower age level.

The following stages may be selected in formation of Stanovoy ophiolites: A - formation of dike and sill complexes in the spreading zones of the marginal basin; B - formation of high-pressure metamorphites (garnet amphibolites) in the subduction zone; C, D - outcrop of ultrabasic and metamorphic plates at a collision of the accretion block with the Siberian continent. In terms of structure and rock composition, the Stanovoy ophiolite belt has common features with similar formations of subduction zones -

Chagan-Uzun belt in Gorny Altai and Indonappu belt in Japan [28]. Only the age of metamorphism of garnet amphibolite protolith and age of their exposure (626-573 Ma) has been established for this belt, These data seem to suggest the boundary age for a obduction of Stanovoy ophiolites and sedimentary-volcanogenic complexes on the margin of the Siberian continent.

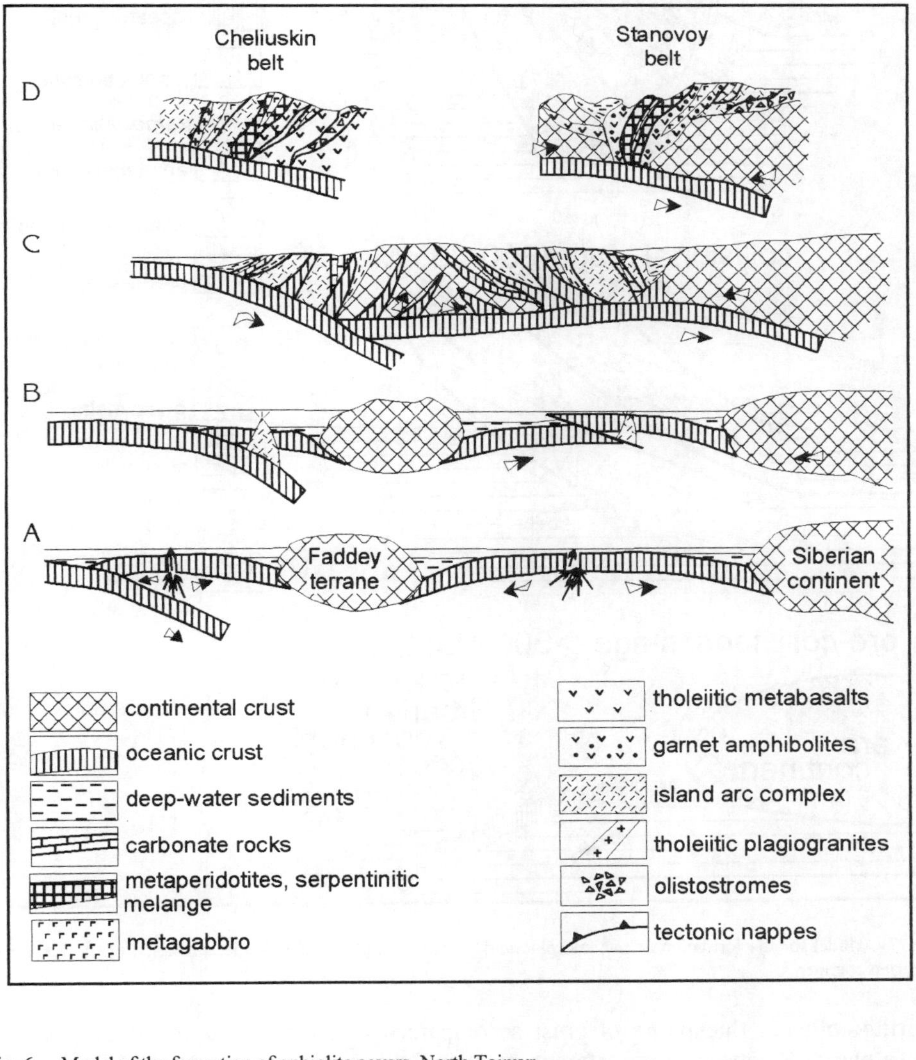

Fig. 6. Model of the formation of ophiolite covers, North Taimyr.

The next large tectonic event in the history of Taimyr is a collision of the Siberian and Kara continents in the Late Paleozoic. The model of formation of granitic and metamorphic complexes in this case suggests three stages: pre-collisional, collisional, and post-collisional (Fig.7).

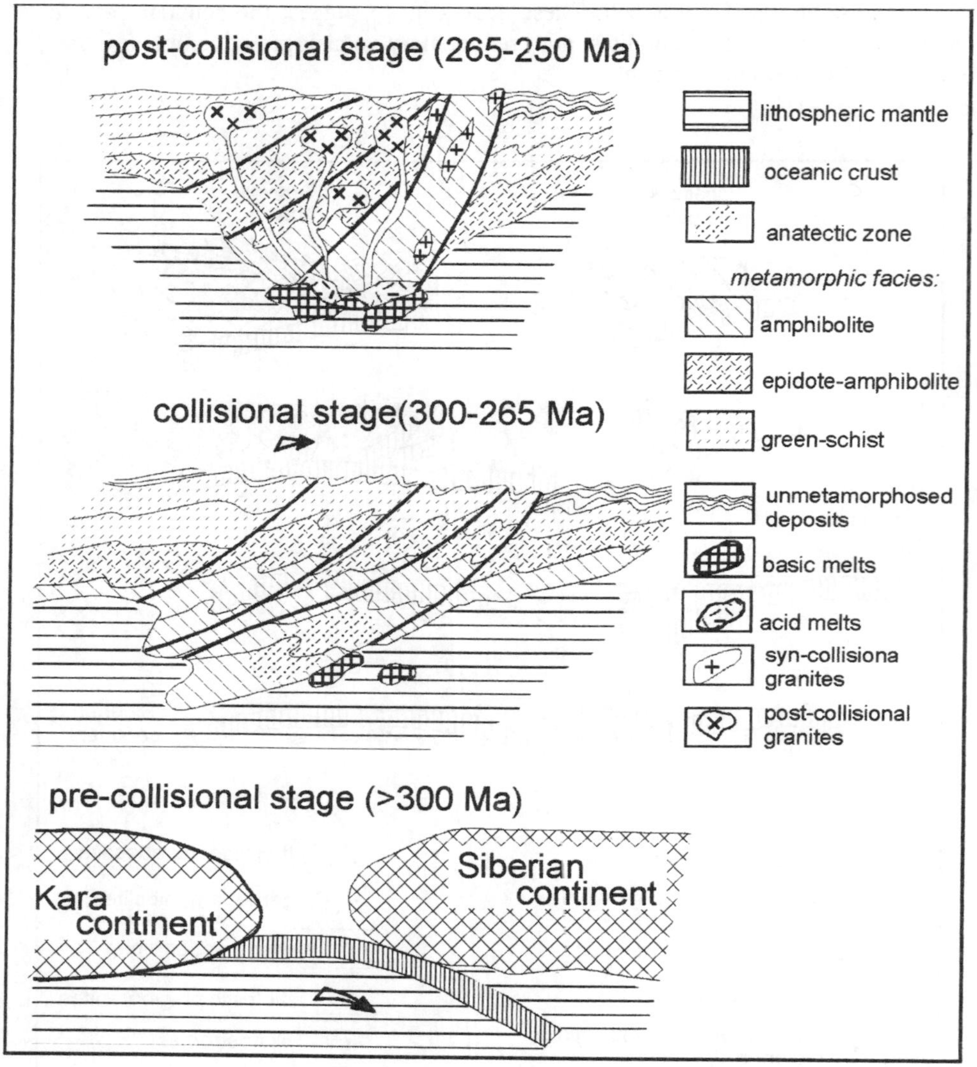

Fig. 7 Model for the formation of metamorphic and granitic complexes during the collisional of the Kara and Siberian continents.

During collision thickening of crust accompanied by folding and thrust dislocations takes place. A zone of ultrametamorphism is formed in the lower part of thickenned crust where the blocks of basic rocks of Neoproterozoic accretion belt and overthrust flyschoid sediments of the Kara continent are present. Here anatectic granites and migmatized metamorphites of moderate pressure amphibolite facies are formed, which are sequentially replaced by middle- and low-temperature zones of regionally metamorphosed rocks in upper horizons. Further thickening of crust due to gravitational unequilibrium leads to "collapse" [28] - quick thrusts, extension and uplift of rock blocks of low crust. The next is postcollisional stage whose specific feature is granitoid fusion in deeper and higher-temperature conditions than granitic anatexis of the preceeding stage. Probably, mixing of hypogene basic magmas of supernatant diapir with acid melts - products of melting of continental crust, plays a significant role in this case.

Age datings obtained for granitoids and metamorphic rocks for each stage (Table 1) reflect the evolution of formation of the Kara orogen. The time interval between formation of syn- and postcollision granitoids is 35-40 Ma. The time of cooling of granite-metamorphic complexes to 500-300 °C is registered due to different temperatures of closing of various isotopic systems. This interval is 30-35 Ma for syncollisional and 10-12 Ma -for postcollisional stages.

DISCUSSION AND CONCLUSIONS

The establishment of the structural and age features of the main magmatic and metamorphic complexes of Taimyr su~ciently clarifies the history of development of this region and its position in the tectonic evolution of the Arctic and Siberian craton, mainly, in the Neoproterozoic and Late Paleozoic.

Currently available geological and geochronological data on the Precambrian complexes of the Arctic structures [29-33], as well as the paleoreconstructions of localization of the continents and ancient oceans [34-39] suggest that in Neoproterozoic there existed the collision-accretion belts of Grenvillian or slightly younger age between the Lavrentia with Greenland, Siberia, Baltica and Barentsia. The presence of such Grenvillian belt between the Eastern Greenland, northern part of the Ellesmere Island from one side, and Baltica with Svalbard, from the other hand, was repeatedly described but its extension on the periphery of Siberia (in a direction of the Angara fold belt on the one hand and the Franklinian fold belt - on the other) either was not discussed because of the absence of data on the Siberian belts or was only anticipated.

Geological and geochronological data show that the range of established Neoproterozoic collisional-accretional events in the Arctic structures is wide enough (1000-700 Ma) and, probably, includes both the events of Grenvillian orogeny and the events related to collision of island arcs with microcontinents. The break time of desintegration of the Neoproterozoic supercontinent (Rodinia) which was preceeded by a Grenvillian orogeny is also determined by wide age interval. Opening of the Paleoasian ocean Paleo-Pacifica is determined by an interval of 900-750 Ma according to the age of ophiolitic and island-arc complexes obducted on the continents. Opening of the first of the former might have taken place about 900 Ma ago [40]. The U-Pb age data on island-arc volcanites and plagiogranites, dikes and sills of diabases, traced from Canadian Cordeliers through Yukon and Franklin belt of North America, through Alaska-Chukot block to Taimyr (Fig 8.) may be indicative of the time of opening of Northern Paleo-Pacific. The datings of magmatic formations of the above-mentioned belts characterize both time of spreading and subduction processes, and are chiefly restricted to the interval of 700-780 Ma, which allows the opening of the North Paleo-Pacific to be attributed to the boundary of 800 Ma.

The following stage of significant tectonic events in the Arctic begins in the Devonian by collisional processes in Alaska, Pearya (Canadian Archipelago) and Svalbard. They were probably related to the formation of the Euroamerica [1, 3]. According to these authors, Lavrasia has been generally formed by the Early Permian, but up to the Middle Jurassic the South-Anyui oceanic bay of the Paleo-Pacific ocean (Pantalassa) existed, which separated Chukotka and Kara continent from Siberia. As to the Kara block, we disagree with the last conclusion. In this paper we showed the formation of the Late Paleozoic orogen at the place of collision of the Kara continent and Siberia. Closing of the South-Anyui bay in Taimyr began in the Late Paleozoic and continued in the Permian, simultaneously with closing of the Paleoasian ocean [1, 3] and termination of the formation of the Late Paleozoic Pangea.

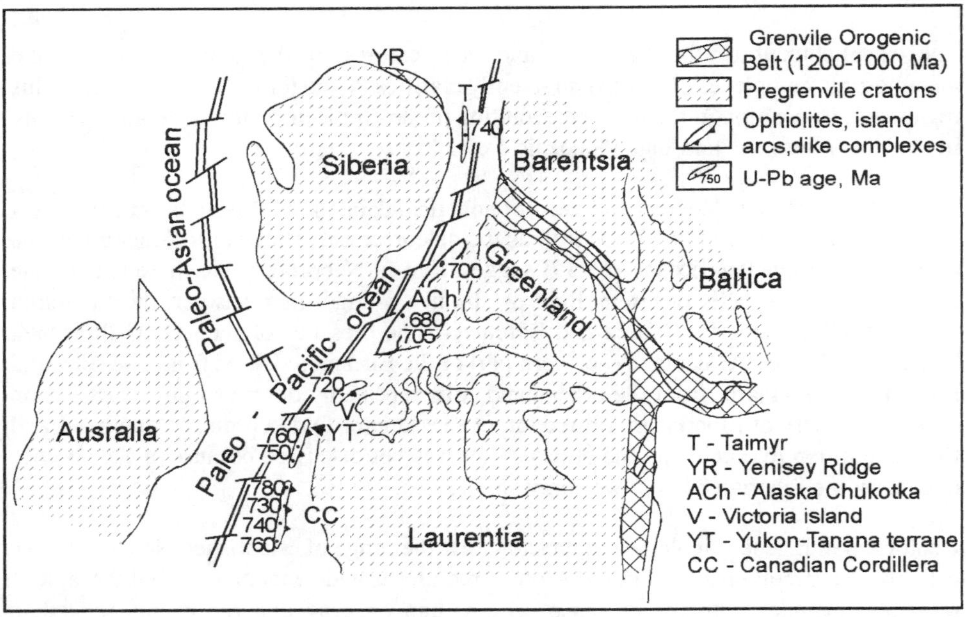

Fig. 8. Proposed paleogeodynamic reconstruction of the Paleo-Pacific ocean opening at 800-700 Ma. Position of Siberia, Greenland, Baltica and Laurentia after Condie and Rosen [38]. Ages of Late Proterozoic island arc and rift-related magmatism (U-Pb ages in Ma) from [1 5, 36, 39, 41-43].

Acknowledgments

The work was supported by grant 95-05-14602 from the Russian Foundation of Basic Research and International Science Foundation (RPGOOO grant). I thank N.L. Dobretsov and Ch.B. Borukaev for their insightful comments, and A.E.Vernikovskaya, A.B. Kotov, B.A. Ponomarchuk, E.B. Sal'nikova, V.P. Kovach for joint geochemical and isotopic-geochronological investigations, and A.V. Radosteva, A.I. Chernykh for graphics and typing.

REFERENCES

1. Zonenshain, L.P., and Natapov, L.M. Tectonic history of Arctics. In: *Actual problems of continent and ocean tectonics, Moscow, Nauka,* pp.31-57 (1987) (in Russian).
2. Uflyand, A.K., Natapov, L.M., Lopatin, V.M., and Chernov, D.V. To the Taimyr tectonic nature. *Geotectonika*, **6,** 76-93 (1991) (in Russian).
3. Zonenshain, L.P., Kuzmin, M.I. and Natapov, L.M.. *Lithospere plate tectonics of the USSA territory.* II.-M., Nedra, 334 p. (1990) (in Russian).
4. Urvantsev, N.N. *The Taimyr folded zone.* Bull. Norilsk. kombinata, Norilsk, pp. 4-12 (1949) (in Russian).
5. Ravich, M.G. The Precambrian of Taimyr. *Tr. NIIGA, Leningrad,* **76,** 312 (1954) (in Russian).
6. Pogrebitsky, Y.E. *Paleotectonic Analysis of the Taimyr Fold System.* Tr. NIIGA, Leningrad, Nedra, **166**, 284 (1971) (in Russian).
7. Bezzubtsev, V.V., and Zalyaleev, R.Sh., and Sakovich, A.B. *Geological map of the Gorny Taimyr 1 : 500000.* Explanatory note, Krasnoyarsk, 177 p. (1986) (in Russian).
8. Zabiyaka, A.I., Zabiyaka, I.D., Vernikovsky, V.A., et. al. *Geological structure and tectonic evolution of northeastern Taimyr.* Novosibirsk, Nauka, 144 p. (1986) (in Russian).
9. Vernikovsky, V.A. Metamorphic formation and geodynamics of Northern Taimyr. *Russian Geology and Geophysics*, **33**,.42-49 (1992).
10. Vernikovsky, V.A. To geodynamic evolution of the Taimyr folded area. *Proceedings ICAM, Magadan, Russia*, Sept.1994, pp.186-193 (1995a).
11. Vernikovsky, V.A Riphean and Paleozoic metamorphic complexes of the Taimyr fold belt: Conditions of Formation. *Petrology*, **3:1,** 55-72 (1995b).
12. Makhlayev, L.V. Tectonic nature of the Mamont-Shrenk block, Central Taimyr. *Izv. Ac. Nauk SSSR, Ser. geol.*, **4,** 77-87 (1988) (in Russian).
13. Vernikovsky, V.A, Vernikovskaya, A.E., Chernykh, A.I., and Melgunov, M.S. Petrology and geochemistry of Taimyr Riphean ophiolites. *Russian Geology and Geophysics*, **36:1,** 103-120 (1996).
14. Coleman, R.G., Peterman, Z.E. Oceanic plagiogranites. *J.Geophys. Res.*, **80,** 1099-1108 (1975).
15. Vernikovsky, V.A., Vernikovskaya, A.E., Lyapunov, S.M. et. al. (1994). Petrology, Geochemistry, and Tectonic Settings of Plagiogranites of the Cheliuskin Ophiolite Belt. Intern. *Geol. Rev.*, **36,** 961-974.
16. Vernikovsky, V.A., Neimark, L.A., Ponomarchuk, V.A., et.al. Geochemistry and Age of Collision granitoids and Metamorphites of the Kara microcontinent (Northern Taimyr). *Russian Geology and Geophysics*, **36,** 46-60 (1995c).
17. Kuzmin, M.I. *Geochemistry of magmatic rocks of Phanerozoic mobile belts.* Novosibirsk, Nauka. 200 p. (1985) (in Russian).
18. Zakharov, Y.I., Ravich, M.G., and Shulyatin, O.G. Metamorphic facies of the Taimyr folded area. In: *Asian metamorphic complexes, Novosibirsk, Nauka*, pp.164-176 (1977) (in Russian).
19. Khain, V.E. Nonalpine Asia and Australia. *Regional Geotectonics,* Moscow, Nedra, **3,** 356 (1979) (in Russian).
20. Zakharov, Y.I., Chukhonin, AP., and Proskurin, V.F. New isotope geochronological data for granitoids of the Shrenk-Mamont projection of the Taimyr peninsula. *Dokl. RAN*, **332,** 58-61, (1993) (in Russian).
21. Makhlaev, L.V., Andreichev, V.L., Golubeva, I.I, et.al. Rb-Sr Dating of Cristaline Rocks in the Mamont-Shrenk block, Central Taimyr. *Dokl. RAN*, **324,** 1265-1269 (1992) (in Russian).
22. Vernikovsky, V.A, Sal'nikova, E.V., Kotov, A.V., et. al. Precambrian granites of the Faddey terrane (North Taimyr): new geochemical and geochronologic (U-Pb, Sm-Nd) data. *Dokl. RAN*, in press, (1997) (in Russian).
23. Vernikovsky, V.A., Kotov, A.V., Ponomarchuk, V.A. et al. Late Riphean-Vendian history in Taimyr: Sm-Nd, Rb-Sr, K-Ar Age of Garnet amphibolites of the Stanovoy ophiolite belt. *Dokl. RAN*, in press, (1997) (in Russian).
24. Vernikovsky, V.A., Sal'nikova, E.V., Kotov, A.V., et. al. Age of postcollision granitoids, North Taimyr: U-Pb, Sm-Nd, Rb-Sr and Ar-Ar data. *Dokl. RAN*, in press, (1997) (in Russian).
25. Pearce, J.A., Lippard, S.J., Roberts, S. Characteristics and tectonic significance of supra-subduction zone ophiolites. In: B.P.Kokelaar and M.F.Howells (Editords), Marginal Basin Geology, *Geol. Soc. London, Spec. Publ.*, **16,** 77-96 (1984).
26. Ewart, A., Brothers, R.N., Mateen, A. An outline of geology and geochemistry and possible petrogenetic evolution of the volcanic rocks of the Tonga-Kermadec-New Zeland island arc. *J. Volcanol. and Geochetm. Res.*, **2:3,** 205-250 (1977).
27. Sharaskin, A.Y., Pusthin, I.K., Zlobin, S.K., and Kolesov, G.M. Two ophiolite sequences from the basiment of the northern Tonga arc. *Ophioliti*, **8,** 411-430 (1983).

2S. Dobretsov, N.L. Tectonics and Metamorphism: Problems of Relationship. *Petrology*, **3 (1),** 4-23 (1995).

29. Peucat, J.J., Ohta, Y., Gee, D.G., and Bernard-Griffiths, J. U-Pb, Sr and Nd evidence for Grenville and latest Proterozoic tectonothermal activity in the Spitsbergen Calesonides, Arctic Ocean. *Lithos*, **22,** 275-285 (1989).

30. Nelson, B.K., Nelson, S.W., and Till, A.B. Nd- and Sr-isotope evidence for Proterozoic and Paleozoic crustal evolution in the Brooks Range, northern Alaska. *J. of Geol.* **101,** 435-450 (1993).
31. Ohta, Y. Caledonian and Precambrian history in Svalbard: a review, and an implication of escape tectonics. *Tectonophysics*, **231,** 183-194 (1994).
32. Gee, D.G., Bjorklund, L and Stolen, L.-K. Early Proterozoic basement in Ny Friesland-implications for the Caledonian tectonics of Svalbard. *Tectonophysics*, **231,** 171-182 (1994).
33. Gee, D.G., Johansson, A., Ohta, Y., et al. Grenvillian basement and a major unconformity within the Caledonides of Nordaustlandet, Svalbard. *Prec. Research*, **70,** 215-234 (1995).
34. Dalziel, I.W.D. Pacific margins of Laurentia and East Antarctica-Australia as a conjugate rift pair: Evidence and implications for an Eocambrian supercontinent. *Geology*, **19,** 598-601 (1991).
35. Hoffman, P.F. Did the breakout of Laurentia turn Gondwanaland inside-out? *Science*, **252,** 1409-1412 (1990).
36. Heaman, L.M., LeCheminant, A.N., Rainbird, R.H. Nature and timing of Franklin igneous events, Canada: Implications for a Late Proterozoic mantle plume and the breakup of Laurentia. *Earth and Planet. Sci. Letter*, **109,** 117-131 (1992).
37. Powell, C.Mc.A., Li. Z.X., McElhinny, M.W., et al.. Paleomagnetic constraints on the Neoproterozoic breakup of Rodinia and the mid-Cambrian formation of Gondwana. *Geology*, **21,** 889-892 (1993).
38. Condie, K.C., Rosen, O.M. Laurentia-Siberia connection revisited, *Geology*, **22,** 168-170 (1994).
39. Patrick, B.E., McClelland, W.S. Late Proterozoic granitic magmatism on Seward Peninsula and a Barentian origin for arctic Alaska-Chukotka. *Geology*, **23,** 81-84 (1995).
40. Dobretsov N.L., Berzin N.A., Buslov M.M. Opening and tectonic evolution of the Paleo-Asian ocean. *Intern. Geol. Rev.*, **37:4,** 335-360 (1995).
41. Roots, C.F., Parish, R.R. Age of the Mount Harper volcanic complex, southern Ogilvie Mountains, Yukon. In: *Radiogenic age and isotopic spudies, Geological Survey of Canada,* Paper 88-2, pp.29-35 (1988).
42. Hoffman, P.F. Precambrian geology and tectonic history of North America. In Bally, A.W.Palmer, A.R., eds., The geology of North America: An overview: Boulder, Colorado, *Geological Society of America, Geology of North America*, **A,** 447-512 (1989).
43. Cecile, M.P., Harrison, J.C., Kos'ko, M.K., and Parrish, R.R. Precambrian U-Pb ages of igneous rock, Wrangel Complex, Wrangel Island, USSR. Can. *J. of Earth Sci.*, **28,** 1340-1348 (1991).

Proc. 30th Int'l Geol. Congr., Vol.7, pp. 139-155
Xu Zhiqin, Ren Yufeng and Qiu Xiaoping (Eds.)

Phanerozoic Domains in a New International Tectonic Map of Europe

V.E. KHAIN, YU.G. LEONOV, V.D. BREZHNEV, M.L. KOPP, A.V. RJAZANTSEV, A.L. VASSILCHOUK

A new International Tectonic Map of Europe, scale 1:5.000.000, has been compiled by the Subcommission of Tectonic Maps of the Commission for the Geological Map of the World (CGMW). The Map is based on the materials (mostly regional drafts) provided by authors from most European countries. They were edited and put together by the Map's Editorial Board, Editors-in-chief V.E. Khain and Yu.G. Leonov, and the team from the Geological Institute and Commission for International Tectonic Maps of the Russian Academy of Sciences. Besides the European continent, the Map covers the territory of East Mediterranean countries, North Africa, North Atlantic and North European shelves. The compiled Map differs from previous editions not only in scale (two earlier editions have been compiled in scale 1:2.500.000)but by a number of other modifications. The main of them are:

(1) Another ideology of the Legend: the main principle of subdivision according to the age of deformations and other (correlated with them) tectonic events is nearly the same. But geodynamic setting of rock complexes, mainly in principles of plate tectonics, has been added.

(2) Far more detailed representation of the Atlantic ocean floor tectonics and the Mediterranean, Black sea and Caspian sea tectonics.

(3) Important modifications in subdivisions and correlation of Precambrian rocks especially for shields and basement of the East European platform.

(4) Compilation on the base of up-to-date information of the inset map, which represents the basement structure of the Russian platform (shown also in the previous edition) as well as of Epi-Paleozoic platforms of Central and Western Europe.

(5) Far more detailed subdivision of platform covers into complexes, that corresponding with the main stages of platform development history. The adopted style of representation of covers gives to some extent the possibility to correlate stages of development of continents and oceans, of platforms and fold belts.

(6) More detailed and up-to-date subdivision of shelves, of sediments and volcanoes of continental slope and continental rise.

We consider here several features of the European Phanerozoic tectonics, which the new Map presents in more detail and interprets differently as compared to the above mentioned two earlier editions. The features to be discussed are the following:

- outer boundaries and distribution of terranes and sutures within the area of Paleozoic orogeny of Central and Western Europe;
- remobilized pre-Paleozoic complexes in European Paleozoides;
- rifting and spreading in North Atlantic;
- volcanism and alteration of the crust at the Atlantic passive margin of North-West Europe;
- some peculiarities of tectonics of Mediterranean seas;
- intraplate deformation.

OUTER BOUNDARIES AND DISTRIBUTION OF TERRANES AND SUTURES WITHIN THE AREA OF PALEOZOIC OROGENY OF CENTRAL AND WESTERN EUROPE.

In addition to the main Map the structure of Paleozoic fold belt is shown on the inset map of 1:10,000,000 scale, which presents basement with a removed Meso-Cenozoic platform cover. Paleozoic complexes are represented on it schematically, but all the same in a rather clear way, which makes easily seen their main structural features. Most interesting is: to trace the outer boundaries of the Variscan belt, to identify and correlate terranes and sutures within its limits, to stress the character of bilateral symmetry of Variscides. A good amount of literature has been used during this work. Publications providing the general subdivision and correlation of Variscan terranes [4, 11, 13, 21, 44] must be mentioned at the first place.

The northern Variscan deformation front does not coincide with the front of Caledonian deformations (Fig. 1). The Variscan deformation front lies considerably to the north. The Caledonian front limits the outer Caledonides of the London-Brabant massif and extends southward for several dozens of kilometers, where it subsides beneath the overthrusting Variscan complexes of the Rhenohercynian zone. The Caledonian basement outcrops here from beneath the Variscides in the massifs of Ardennes and the Rheinische Schiefergebirge.

The problem of the Variscan belt southern boundary deserves analysis which is beyond the scope of this paper. Definition of its position is very complicated, because on one hand, it is overlapped by Alpine complexes, and on the other, is camouflaged by the crustal destruction in the course of formation of Mediterranean Neogene basins. At any case, the Pennine complexes of Alps belong to the Variscan belt: within Pennine basin, probably since Ordovician, the structures of the south-east margin of the Early Paleozoic Ocean were formed. They have been preserved in nappes of North Galicia, South Armorica, in Central Massif and Moldanubicum [43]. In recent structure the Variscan basement of Alps represents a part of the Variscan belt southern flank. The available data permit to place the southern Variscan deformation front even more southward, outside the present-day Peri-Adriatic line. At the same time this front acts as the northern boundary of distribution of complexes of the passive margin of Gondwana's Adriatic plate [26, 62]. As a whole, the outer boundary of the southern branch of West European Variscides can be rather tentatively traced south of Precambrian and Paleozoic complexes of Austrian and Southern Alps, east of Corsica and Sardinia and west of the Adriatic plate. The southern Variscides include also tectonic nappes of Calabria and

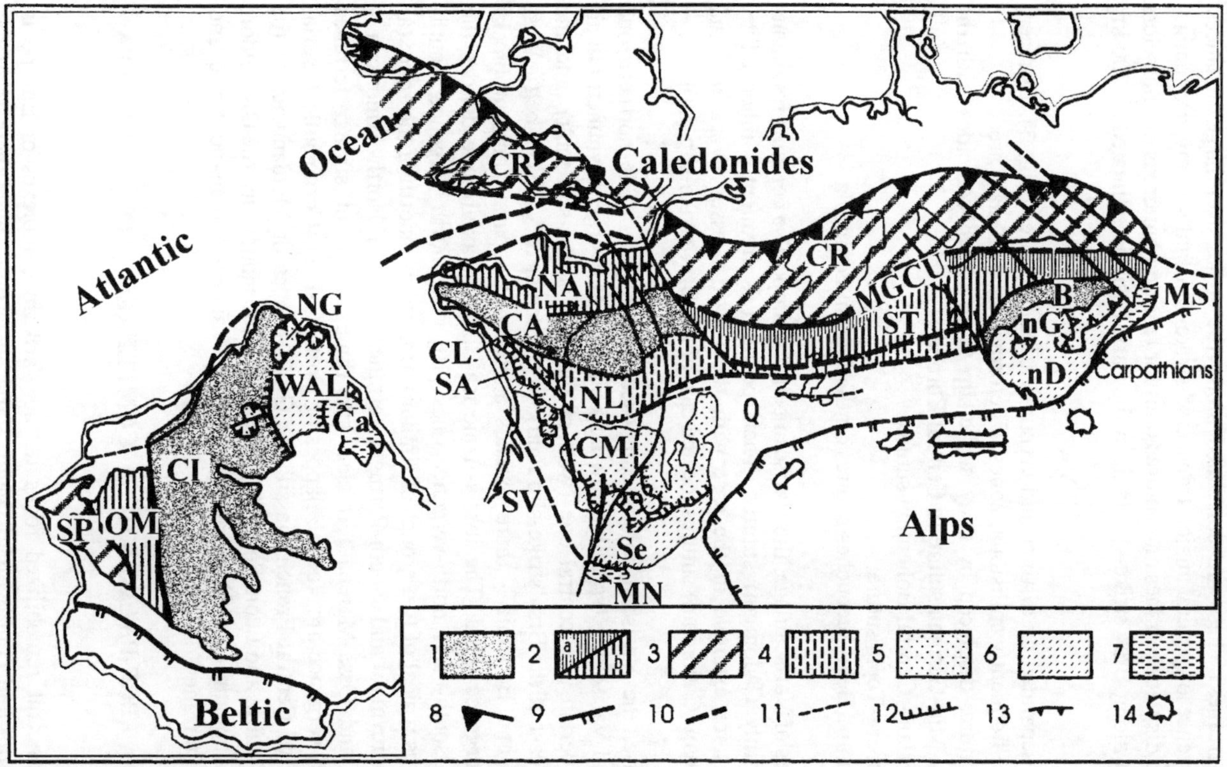

Fig.1. Scheme of terranes of the West European Variscan Belt.
1-Central Iberian terrane with North Iberian parautochthon (Cl), Central Armorica (CA) and Bohemian (B) axial terranes (with Cadomian basement and Variscan deformed cover). Terranes with the western and northern vergence: 2-Ossa-Morena (OM), Northern Armorica (NA), Mid-German cristalline uplift (MGCU) - a, Saxo-Thuringian (ST) - b, (with Precambrian blocks); 3-South-Portugal (SP) and Cornwall-Rhenohercynian (CR). Terranes with the eastern and southern vergence: 4-North Ligerian margin of Armorica; 5-Nappes of North Galicia (NG), Cornouaille-Ligerian (CL), the northern part of Central Massif (CM) and nappes Gfohl (nG) of Moldanubicum (with the Precambrian blocks); 6-West Asturia and Leon (WAL), South Armorica (SA), Sevenne Mountains (Se), and nappes Drosendorf (nD) of Moldanubicum; 7-Cantabrian (Ca), South Vandee (SV), Montagne Noire (MN), Moravian-Silesian (MS); 8-Variscan deformation front; 9-Alpine deformation front; 10-sutures; 11-other faults including strike-slips; 12-thrusts; 13-nappes; 14-tectonic windows in Alps.

massifs of Lesser and Greater Cabjilie in North Africa [10, 54].

In contrast to accretional British Caledonides, the Variscan collision belt of West Europe has been formed as a result of interaction of ancient blocks within the northern continental margin of Gondwana [44]. The belt was formed by a complicated combination of terranes. These are correlated along the strike over large distances, despite the occurrence of several major transverse strike-slip faults (Bray, Bohemian, Seine etc.), which displace terranes and suture zones for tens and hundreds of kilometers. The displacement along Bray strike-slip fault in the Paris basin is estimated as 200 km (Fig. 2).

The terranes are separated by distinct and cryptic Variscan sutures (Fig. 3). In recent structure these are frequently collisional zones of deformation. The main are:
Lizard-Giessen suture, displaced by Bray strike-slip fault, which coincides with the Cadomian Ushant-Mid-German suture in Central Europe;
Cornouaille- Ligerian - Central Bohemian suture;
Ushant(?)-Münchberg-Tepla suture;
Coimbra-Cordoba and Beja-Acebuches sutures on the Iberian Peninsula.

Besides the sutures of Variscan age, the Variscan belt of Europe also contains fragments of more old (mainly Cadomian) sutures. Their position is frequently inherited by Variscan ones, as for example in the case with the Lizard-Giessen (Variscan) and Ushant-Mid-German (Cadomian) sutures located east of the Bray strike-slip fault.

In the broad sense, the Variscan belt is composed of two parts. The northern one comprises classical Variscides, typically showing a bilateral antivergent symmetry in all their segments: Iberian, the central one (west of the Bray strike-slip fault), and the eastern segment (east of the Bray strike-slip fault). In the center of Variscan belt - West Iberia, Armorica and Bohemia - there extends a chain of separate ancient blocks controlling the vergence trend. The Iberia-Armorica-Bohemia axial zone divides the terranes into the south- and north-vergence ones [10]. The north vergence (south-western in Iberia) is typical of two groups of similar terranes lying north of the zone: (a) Cornwall-Rhenohercynian and South-Portugal terranes and (b) North-Armorican - Saxo-Thuringian and Ossa-Morena terranes. The three groups of similar terranes showing the south vergence are: (a) the nappes of North Galicia, Cornouaille-Ligeria, the northern part of Central French massif; the Gfohl nappes of Moldanubicum; (b) terranes of West Asturia and Leon, South Armorica, Sevenne mountains, the Drosendorf nappes of Moldanubicum; (c) the Cantabrian, South Vandee, Montagne Noire and Moravian-Silesian terranes.

REMOBILIZED PRE-PALEOZOIC COMPLEXES IN EUROPEAN PALEOZOIDES

The problem of ancient remobilized rocks in the Variscan complexes of Europe is extremely interesting and controversial. Many Variscan terranes have their basement composed with Late Precambrian, mostly Cadomian, complexes having suffered both the Cadomian and intensive Variscan tectono-metamorphic reworking and granitization.

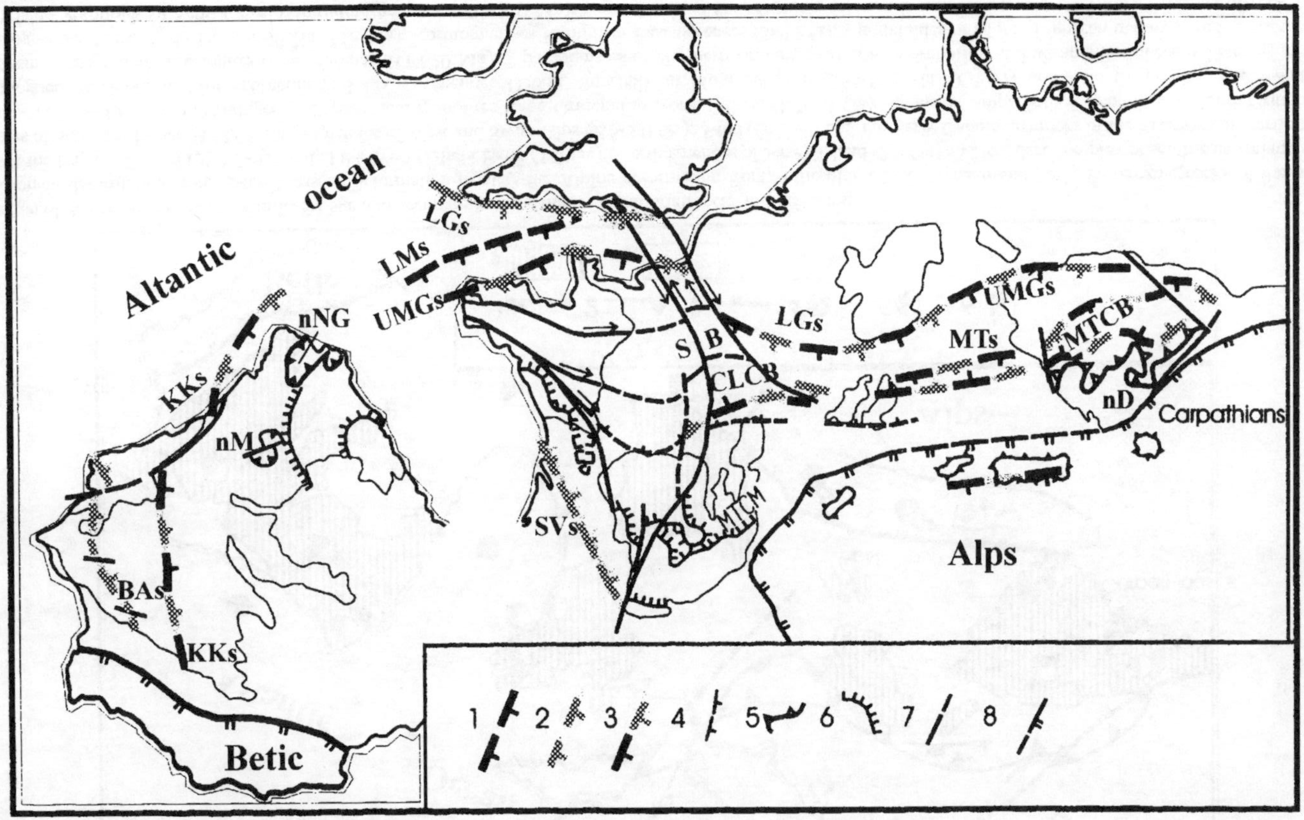

Fig.2. Scheme of sutures of the West European Variscan Belt.
1-Cadomian (or older) suture La Manche (LMs); 2-Variscan sutures with mafic-ultramafic rocks: Beja-Acebuches (BAs); Lizardian part of Lizard-Giessen (LGs); Munchberg-Tepla (MTs), Southern Vandee (SVs); 3-complex of Cadomian-Variscan sutures: Koimbra-Kordoba (KKs), Ushant-Mid-German (UMGs), Main Thrust Central Bohemian (MTCB), Munchberg-Tepla (MTs); 4-Variscan rootless sutures: in the nappes of North Galicia and Morais (nNG and nM), in the Drosendorf nappes (nD); 5-Variscan "dry sutures": Ligerian-Central Bohemian (CLCB), Main Thrust of Central Massif (MTCM); 6-thrusts; 7-Other faults including strike-slips (Bray - B, Seine – S); 8-Alpine deformation front.

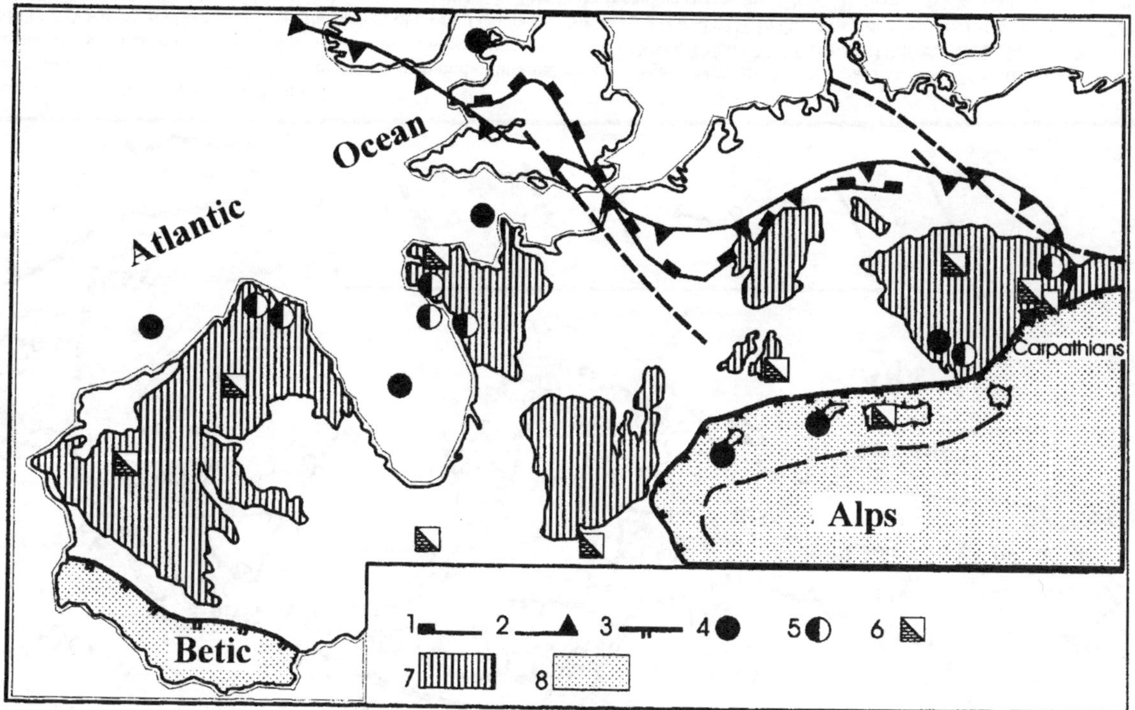

Fig.3. Position of deformational fronts of different age and occupation of the main Precambrian rocks in Variscides:
1-the Caledonian deformation front; 2-the Variscan deformation front; 3-the Alpine deformation front; 4-the oldest Late Archean and Early Proterozoic rocks of the basement: - granulites of the Bay of Biscay (1,9-2,7 Ga) and of the West Galicia Bank (1,5 Ga); – orthogneisses of Jersey Island (2,0 Ga) and Rosslare complex of south-east Ireland (2,1 Ga); - orthogneisses of Southern Bohemia (2,05 Ga.) - gneisses of West and Swiss Alps (2,5-3,0 Ga); 5-old (Grenvill and Pre-Early Cadomian) rocks of the basement, underwent Variscan metamorphism (480-490 Ma and younger): - gneisses and granulites Kabo Orthegal in North Galicia (1,2-1,3 Ga); - gneisses, amphibolites, granulites of South Brittany (1,2-1,3 Ga); - orthogneisses Dobra in Moldanubicum (1,3 Ga.) – gneisses Deblin, Snezhnik in Moldanubicum (565-560 Ma.) 6-Early and Late Cadomides, underwent Variscan metamorphism: – amphibolites, eclogites Ossa-Morena (611-620 Ma); – paragneisses of Northern Brittany (801 Ma); - granulites and ultramafites of Saxo-Thuringiari (705~828 Ma); - orthogneisses Mirando do Duro in Iberia (618 Ma); - orthogneisses Kanigu in East Pyrenees (580 Ma); - granitoides of More, Tanneron massives and in Montagne Noire (600-530 Ma); - granitoids of Schwarzwald, intruding gneisses (530-520 Ma); – granitoids of the Moravian-Silesian terrane's basement (584 Ma); 7-Limits of rock outcrops in Variscides of Western Europe; 8-Alpine-Carpathian area.

These processes camouflage the Cadomian complexes infrastructure, thus creating the problem of subdivision of metamorphosed Paleozoic and pre-Paleozoic rocks [11].

However over the last years, there has become a tradition to ignore the indications on the pre-Paleozoic isotopic age of certain rocks within Variscides, especially obtained by zircons. We may refer to numerous pre-Paleozoic zircon dating, including those when one and the same rock complex yielded ancient ages (1.2 - 1.3 Ga and more) together with Variscan isotopic age of metamorphism (Fig. 1). Such cases are particularly typical for the western parts of Variscan belt. Some of them are: granulites dated as 1.5 Ga in West Galicia bank (the Vigo uplift) [13]; dating of some 1.3 Ga, possibly reflecting the age of MORB protolith in the rocks of Kabo Orthegal synform, Northern Spain (for which Variscan date of 490- 480 Ga has also been established as the age of metamorphism) [6, 7]. Similar situation exists in South Brittany where the age of gabbro protolith in the Dezessar horst has been reported as 1.3 Ga in spite of the age of metamorphism of some 435 Ma; the granulites dated as 1.9-2.7 Ga were found on the floor of the Bay of Biscay; pyroxenes dated as 1.2 Ga from the Odiern Bay in South Brittany with the age of metamorphism 386 Ma [48]; orthogneisses with the age about 2.0 Ga in North Brittany and the Jersey Isl. as well as the Rosslare gneisses in south-east Ireland [14, 50]. In North and South Brittany [48, 49], Saxo-Thuringia [2], Helvetian Alps [23] a level of 800-860 Ma has also been recognized by zircons, U-Pb and Sm-Nd methods. Another –Cadomian (600-660 Ma) - isotope-chronological boundary, important for the Variscides prehistory has recently been established in Iberia [1] and the Pennine nappes of Alps [22].

The above data bring to our attention the problem of relations of pre-Variscan and Variscan complexes with an intricate history of their polyphase deformation, metamorphism and granitization. The results of studies of metamorphic complexes in tectonic nappes of Iberia [42] can be an example of a new approach to this problem. It should be recognized that the extent of participation of remobilized ancient rocks in the structure of Variscan complexes was significantly higher than it is assumed in many of the present-day publications. Since the ancient complexes were most possibly of Gondwana origin, the role of Gondwana basement in the generation of the Variscan belt colission structure was possibly significant.

The distribution pattern of Variscan mini oceanic basins is distinctly inherited from the older ones. For example, a Cadomian Celtic ocean has been inherited almost within the same margins by a Variscan Rheic oceanic basin and, respectively, the eastern part of Lizard-Giessen and La Manche sutures have been distinguished as a Cadomian-Variscan element of long-term life. The Cadomian Ossa-Morena rift also suffered a second opening in the same place in the Ordovician, and subduction in the Sillurian, which has been established for the Coimbra-Cordoba suture complexes. Within the Alps, or more exact within Cadomian Pennine oceanic basin, since Ordovician time the recurrent formation of basins with mafic-ultramafic magmatism began. During the Cadomian epoch, the three above minioceans were probably marginal basins, fringing the Gondwana continent, monolithic at that time. In the course of the following destruction, this continent yielded Variscan complexes of West Iberia and Armorica, the Bohemian and possibly, Ebro-Aquitanian microcontinents, and the Mid-German uplift.

RIFTING AND SPREADING IN NORTH ATLANTICS

For evolution of opening and spreading in northern Atlantic Ocean are characteristic: (1) the asynchronous onset of these processes at different parts; connection of these processes with rifting, which preceded opening and appeared on continental margin episodically later; and (2) the jump-like migration of spreading axes. The latter is most vividly evidenced by repeated jumps of spreading axes as demonstrated on the Map.

A repeated migration of spreading axes is most comprehensively represented in the Norwegian Sea (Fig. 4). East of the marginal high of the Vöring plateau, the rifting started as early as the Paleocene, with the uppermost Cretaceous beds having been subsided for 4 km [17]. Somewhat later, at the time corresponding to the anomalies 24 A and 24 B, i.e. the Early Eocene, northwestern displacement of spreading axis and the opening of Norwegian and Lofoten basins, separating Eurasia from Greenland took place. In the Norwegian basin, the spreading terminated in the latest Eocene [59]. Symmetric anomalies 15 and the extinct axial spreading Aegir Ridge enclosed between them have been documented there. After that, a jump of spreading axis stoped the opening of the Iceland Plateau basin, with the Yan-Mayen microcontinent starting to separate from Greenland [24, 30] - this microcontinent with subaerial basaltic lava, whose extension perhaps occupies the south-east part of Iceland is for the first time delineated on the Map. The Iceland plateau has a complicated structure with a system of not confidently identified magnetic anomalies [12]. Later, in the Miocene, before the 6B anomaly time, the discussed spreading axis had been extinct at this place and jumped northward into the recent Kolbensey Ridge. The rift axes migration is observed in Iceland as well.

Such structures as the Rockall trough, and the Bay of Biscay basin may have resulted from the same type of processes [12, 28, 63]. In both cases, linear magnetic anomalies have been established beneath a thick sedimentary cover. This permitted to suggest opening and generation of the oceanic crust. The Bay of Biscay basin, oriented at right angle to the recent Mid-Atlantic ridge, opened in the Late Cretaceous, synchronously with the Labrador basin.

Older extinct spreading axes have also been traced on the Map, outside the above-discussed northern sector of Atlantic. The oldest axis, is marked by the Early Cretaceous M10 anomaly. It is located in the area of the Tagus basin, to the west of Iberian Peninsula [40, 45].

The above data evidence the uneven rate of the process of crustal extension and opening of North Atlantic, whose main motive was repeated extinction and jumps of spreading axes.

VOLCANISM AND ALTERATION OF THE CRUST AT THE ATLANTIC PASSIVE MAR IN OF NORTH-EAST EUROPE

All up-to-date information evidences the complicated process of crustal transformation at the Atlantic ocean margins. The leading, or at least the most clearly manifested, role in this process is given to volcanism combined with thinning and reworking of the crust.

Margins of volcanic as well as of non-volcanic type occur in Northern Atlantic [63]. The North Atlantic volcanic province extends for nearly 2000 km in longitude (see Fig. 4). The history of volcanism of the central and marginal parts of this province differs significantly: in the center of the ocean (in Iceland) it extended over 60 Ma, in comparison with it volcanism on margins was rather episodic – from 3 to 5 Ma. Intense volcanism of such type is observed in many places. It is represented by subaerial basalt, usually expressed on seismic profiles by seaward-dipping reflectors. The marginal high of the Vöring plateau, Faeroe-Shetland marginal high, the Eastern Greenland, the western part of Rockall plateau and some others are typical in this respect [18, 19, 53]. Marginal volcanic highs are traced as wide belts, interrupted by transform zones with significant shift. The volcanic formation preceded the rupture of the continental crust.

Petrochemical subdivision of volcanic series into two parts is typical. For instance, on the Vöring plateau, where the maximum of volcanic activity fall within 55-59 Ma., the lower part has a mixed composition: basaltic andesites of MORB type, dacites. The origin of the latter is explained (from trace elements) as due to melting of metasedimentary Caledonian rocks [17, 18]. The upper part is similar to Holocene magmatic rocks of Mohns and Knipovitch Ridges [47] and partly of Reykjanes Ridge [58]. These differences in compositions of the lower and upper parts of volcanic series reflect transition of melting out from continental lithosphere to melting out from oceanic asthenosphere.

Volcanoes at the outer part of marginal highs form secondary inclined seaward- dipping reflectors [59] (Fig. 5). Volcanoes within marginal highs occur in different structural position. Main part of them on the south-eastern Atlantic margin is below the shelf boundary and beyond the foot of continental slope within submerged plateaus and microcontinents. On the western margin of Greenland the main part of volcanoes occur within shelf. This can reflect different conditions of opening of Lofoten and Norwegian basins, which was followed by sequential jumps of the spreading axis into the Greenland basin. In the process, the separation of the Yan Mayen microcontinent happened. In the recent structure it is in the center of North Atlantic, and probably, it is not accidentally that part of Iceland is located on its extension. The possibility should not be excluded that the part of Iceland volcanic rocks [51] was formed in the continental margin conditions – the same as the Vöring plateau.

Towards the continents the marginal highs give way, as a rule, to troughs complicated by normal faults, listric faults and rifts (see Fig. 4,5). The Vöring plateau and Faeroe-Shetland highs are limited by escarps, outside which the volcanic series subsided to different depth. Post-Eocene subsidence is evaluated in 4 km. There, as well, occur more earlier rift structures. Thus, the uplift of the Vöring plateau was accompanied by developing of a rift trough, in which the roof of Cretaceous was submereged for 4 km by the Early Eocene [59].

The western margin of the Barents Sea shelf and Svalbard has several characteristics, reflecting the specificity of its geodynamic situation. The transition zone from

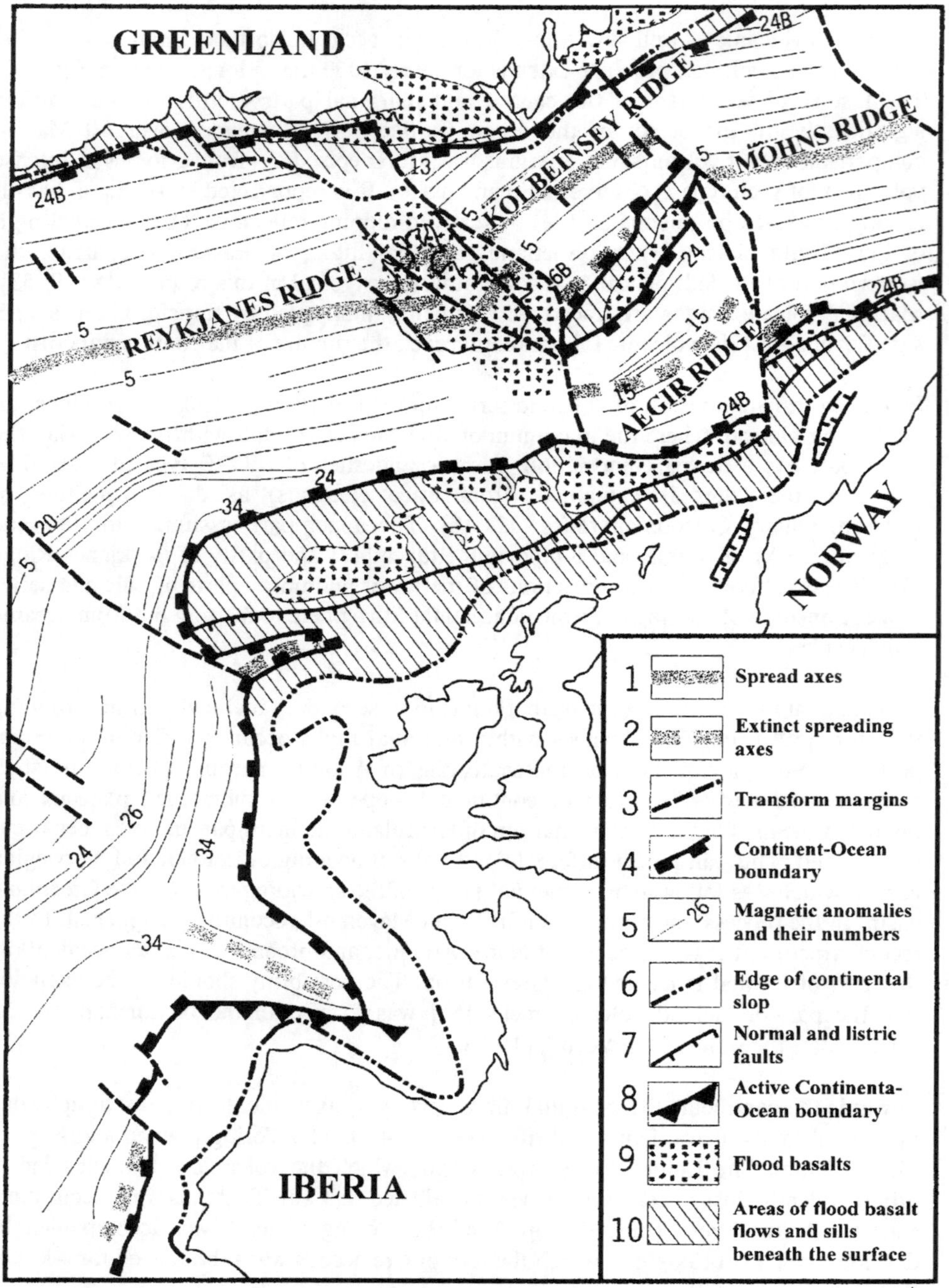

Fig. 4 Scheme of tectonic evolution

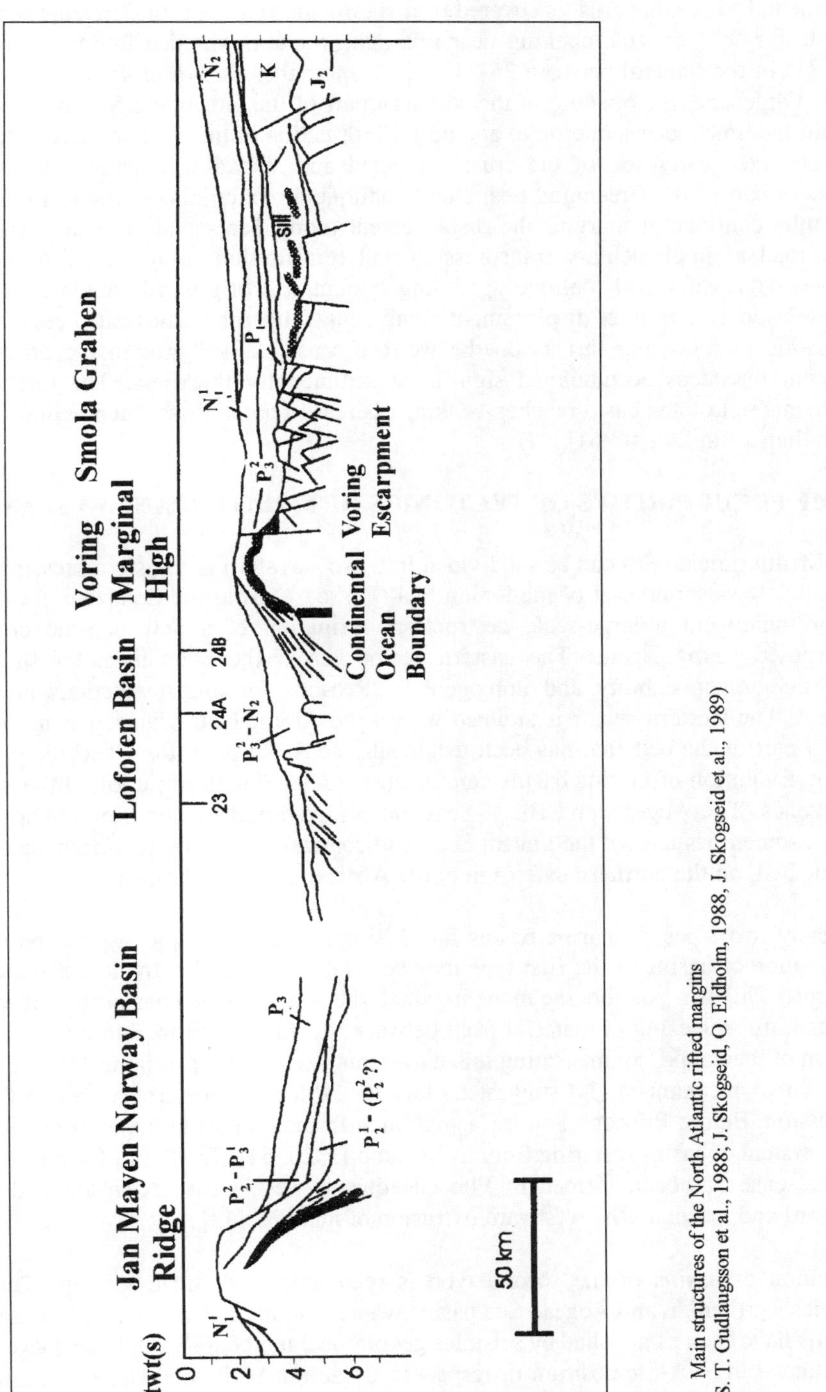

Fig. 5 Main structures of the North Atlantic rifted margins
(from S. T. Gudlaugsson et al., 1988; J. Skogseid, O. Eldholm, 1988, J. Skogseid et al., 1989)

continental to oceanic crust is covered by a significant thickness of Cenozoic sediments. The history of Cenozoic opening near this margin can be divided into two stages [19, 29]: (1) in the interval between 25-24 and 13 anomalies, i.e. from the Early Eocene to Early Oligocene, the opening of the southern part of the Greenland Sea took place, (2) within the time, corresponding to anomaly 13, direction of the relative plate movements has changed, extension of the crust happened and spreading expanded towards the northern part of the Greenland Sea. Due to oblique direction of opening against broken by faults continental margin, the displacement along Senja and Hornsund transform faults had a supplementary transpression and transtension components. A diamond-shaped depression with oblique spreading system has originated in the situation of transtension as a result of displacement along Senja fault zone. The transpression caused formation of folds and thrusts on the western coast of Svalbard. In the process, the adjacent foredeeps accumulated significant sedimentary thickness. The thickness of sediments in Lofoten basin reaches 4-6 km, whereas in other abyssal depressions it is not more than 1 km [28, 46, 61].

SOME PECULIARITIES OF TECTONICS OF MEDITERRANEAN SEAS

The Mediterranean Sea can be subdivided into two - western and eastern sectors, located respectively west and east of the lonian Sea (Fig. 6). During the Cenozoic, the western sector underwent a large-scale destruction, formation of a new oceanic crust and superposed young basins. The eastern sector lacked the phenomena of such type, showing greater stability and homogeneity. Reasons for this distinctions are rather evident. The western sector is situated within the Alpine belt. Marine basins compose here a part of the belt that has been technically active through the Cenozoic up to the recent. Evolution of marine basins here is closely related with evolution of the adjacent fold zones. The Aegean and Black Seas are also referred to this group. The eastern sector, sometimes termed the Levant Sea, is in contrast lying almost entirely outside the Alpine belt, on the northern extension of the African-Arabian platform.

Basically two types of marine basins are distinguished within the Alpine folded belt. Origination of basins of the first type may be related to lateral extrusion of material of the crust. This was possibly the mode in which the Alboran basin has been formed – due to westward squeezing of material from between the Africa and Iberian Peninsula [16]. A front of the nappe, compensating this movement, extended far into the Atlantic Ocean. The same mechanism is suggested for the Pliocene-Quaternary South-Caspian depression. Before Pliocene South Caspian area formed a part of a vast back-arc basin and a system of basins resulting from its breakup [3, 20, 64]. The South Caspian basin in proper sense has been formed in Pliocene due to west – east extension and lateral eastward and, particularly, westward extrusion of material [33].

Formation of basins of the second type is related to back-arc extension. This type includes Tyrrhenian and Aegean Sea basins where a number of normal fault-and-graben systems have been established by seismics geophysical method[37, 56]. The above basins are situated in back-arc position in respect to Calabrian-Sicily and Hellenic island arcs which posses all the island-arc properties: an inclined Benioff zone, a zone of calcareous-alkaline volcanism, presence of deep troughs and trenches separating

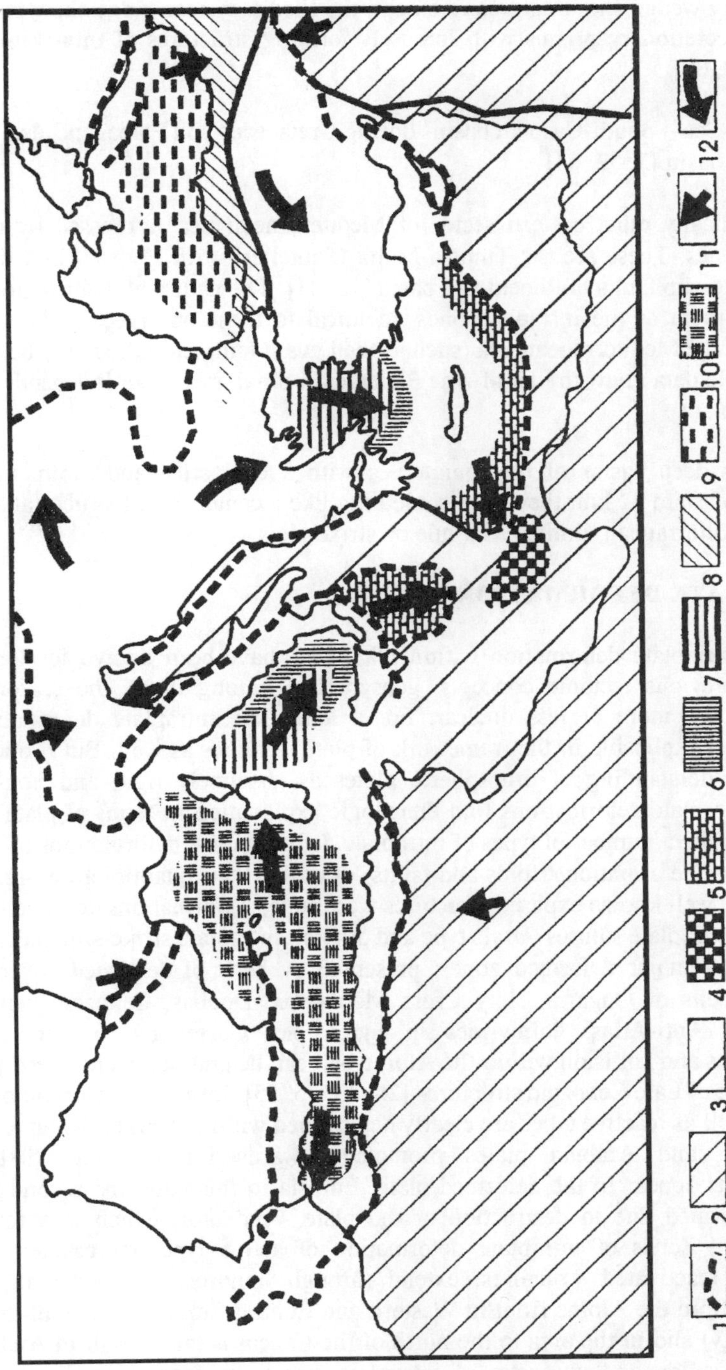

Fig.6. Mediterranean Sea
1 - Alpine front; 2 - African-Arabian plafform; 3 – the same within the aquatory; 4 - Ionic basin; 5 - accretional prism; Lae Cenozoic and recen : 6 - volcanic island arcs, 7 - back-arc basins; Cretaceous-Paleogene: 8 - volcanic island arcs, 9 – back-arc basins; 10 – Extensional basin generated as a result of horizontal extension connected with lateral extrusion of material; 11 – Extra-pressed areas; 12 – Direction of the lateral extrusion.

volcanic arcs and accretionary prisms [8]. The deep-sea Hellenic trench is a compressional structure - a trough split by diagonal strike-slip faults [37, 52]. Messinian sedimentary wedge and, to a greater extent, the Mediterranean ridge show properties of growing accretionary prisms with intensely forming structures of thin-skinned nature [32, 38, 52, 60].

The Black-Sea – South-Caspian basins during Cretaceous and Paleogene also evolved as a back-arc basin [3, 52, 64].

The tectonically quiet eastern sector of Mediterranean Sea is crossed by two active tectonic zones. These are the Tunisia-Malta (Pantelleria) rift [31, 57] and the zone of active strike-slip faults of Phoenician basin [15, 41]. The nature of deformations permits to qualify them as the intraplate ones, identical to those observed in the continental fringing of the Mediterranean Sea (such as the Levant fault, etc.). It should be added that the available data allow to identify the Phoenician-Sinai area as an independent plate or rather subplate.

As regards deep basins of the lonian Sea with a cool crust and sedimentary cover thickness of 10 to 12 km, they form something like a center for all structural arcs of the central Mediterranean, limited as a rule by strike-slips.

INTRAPLATE DEFORMATION

Zones of intraplate deformation within platforms, have been known for long, but in terms of previous tectonic concepts, geosynclinal among them, the reason of their formation was more or less unclear. For a long time intraplate deformations were considered inexplicable in the framework of plate tectonics as well. But if one gives up classical understanding of lithospheric plates as absolutely rigid and non-deformed masses, intraplate deformations find their logic explanation in terms of plate tectonics. There are quite a number of types of intraplate deformations: deformations of basement; local rootless deformations (folds and faults) of cover (their quantity grew significantly, when to the well known exposed structures were added deformations detected by seismic methods); intraplate sutures (Urals type and others); intraplate strike-slip faults and rifts; and at last - major deformed zones, presented by belts of deformed cover, such as Paleozoic belts of Ougarta, Holy Cross Mountains, Donbas, Cenozoic belts of Jura Mountains, Anti-Atlas, Palmyrides in Syria, etc. Correlation between intraplate deformations and collision within the Alpine belt can be and has been indeed proved on the example of Late Cenozoic structures [25, 33, 36, 55]. Intraplate deformations of both active as well as reactive types are clearly determined within zones of dynamic influence of Apulian and Arabian plates promoted towards Eurasia. The first (active) deformations appear in the deformed plate, Eurasia in this case, the second (reactive) ones are formed due to destruction of the plate - indentor, which is a real tool of deformation. Zones of intraplate deformation of active type, represented by newly formed or reactivated structures, extend through hundreds to over a thousand of kilometers from the Alpine front in Western and Central Europe (as a result of Apulian plate activity) and in the area to the north of the Caucasus (as a result of Arabian plate activity). Position and character of intraplate structures depend on parameters of a plate-indentor movement at every certain time span, that determine stress and strain regime of the deforming plate. This is established for Europe [5, 27, 39] as well as for Scythian-Turanian platform [33, 35, 55]. For the latter, the authors show that in the Early Miocene, when the Arabian plate underwent anti-clockwise rotation (while closing of relic Neo-Tethys basin), the western part of the Turanian platform, where the dextral transpression took place, was mostly deformed. By the end of the Late Miocene, when the direction of the Arabian plate movement changed and became nearly meridional, the front of intraplate deformations moved to the west. Meridional compression of the Caucasus and the Scythian platform, and extension, accompanied by formation of the South-Caspian basin in the rear of this compressional belt, happened.

The intraplate deformations of reactive type are especially clearly manifested within the Arabian plate. They are of post Middle Miocene age, when the Arabian plate collided with the Eurasian plate along the collisional suture. The reactive deformations are expressed in crumpling of the basement and cover of the Arabian plate in front of the Zagros-Taurus belt. Due to the lack of place in front of the plate-indentor, folds are forced to bent towards the movement of the later.

Concerning the complicated ensemble of intraplate structures of the west of the Arabian plate: Palmyrides, Levantine (or Dead Sea) fault, and others, the analysis shows that this structural ensemble originated in a strike-slip type stress field. An additional peculiar feature of the situation, which determined formation of the Palmyrides and other intraplate structural elements of this area, is a shape of the Levantine transform fault, forming a curve in Lebanon and blocking the movement of the Arabian plate to the north [34].

Connection of intraplate deformation with horizontal compression resulting from colliding plates can be clearly demonstrated for southwestern Europe and northwestern Africa. Tethys Ocean practically disappears here and area of typical alpine structures becomes narrow: only Rif and Bethic belong to it. But due to horizontal compression deformational front outstepped the area of collisional suture and alpine belt. The territory of Iberian peninsula and northern part of Africa underwent intraplate deformation.

REFERENCE

1. Abalos B., Gil Ibarguchi J.l., Eguiluz L. Cadomian subduction/collision and Variscan transpression in the Badajoz – Cordoba shear belt, SW Spain [BCSB] *Tectonophys.* **199,** 51-72 (1991).
2. Altenberger U., Besch T. The Bollstain Odenwald: evidence for preto Early Variscan plate convergence in the Central European Variscides, *Geol. Rund.* **Bd.82,** 475-488 (1993).
3. *Atlas Tethys. Palaeoenvironmental maps.* J.Dercourt, L.-E.Ricou and B.Vrielynck, eds. Paris, France (1993).
4. Behr H.J., Engel W., Franke W. et al. The Variscan belt in the Central Europe: Main structures, geodinamic implications, open questions, *Tectonophys.* **109,** 15-40 (1984).
5. Bergerat F. Stress fields in the European platform at the time of the Africa-Eurasia collision, *Tectonics,* **6,** 99-132 (1987).
6. Bernhard-Griffiths J., Cornichet J. Origin of eclogites from South Brittany, France: Sm-Nd isotopic and REE study, *Chem. Geol. (Isotope Geosci. Sect.).* **52,** 185-201 (1985).
7. Bernard-Griffiths J., Peucat J.J., Cornichet J. et al. U-Pb, Nd isotope and REE geochemistry in eclogites from Cabo Ortegal complex. Galicia, Spain, *Chem. Geol. (Isotope Geosci. Sect.)* **52:2** (1985).
8. Bocaletti M. et al. The Calabrian arc and the lonian Sea in the dynamic evolution of the Central Mediterranean,

Marine Geology, **5:3/4,** 219-245 (1984).
9. Brezhnev V.D. Tectonic zonation of Variscides of Western Europe. Izv.VUZ. *ser.geol.* **3,** 8-29 (russ.)(1995)
10. Brezhnev V.D. *Terranes and sutures of Variscan Belt of Europe: their Late Precambrian - Early Palaeozoic Evolution.* XIII Peunion de Geologia del Oeste Peninsular.Ann.lGCP Project 319 Meet.Salamanca - Coimbra. P.P.228-230 (1995).
11. Brezhnev V.D. *West European Precambrian basement and its role in the Variscides evolution. Precambrian of Europe: Stratigraphy, Structure, Evolution and Mineralization.* S-Petersburg. P.151 (1995).
12. Cande S.C., LaBrecque J.L., Larson R.L. et al. *Magnetic lineations of the World's Ocean basins.* Lamont-Doherty Geological Observatory of Columbia University Contribution. No 4367. (1989).
13. Capdevila R., Mougenot D. The pre-Mesozoic basement of the western Iberian continental margin and its place in the Variscan Belt. *Proceed. ODP. Sci. 1988. Results.* **103,** 3-12 (1988).
14. Chadwick B.A., Pharaoh T.C., Smith N.J.P. Lower crustal heterogenity beneath Britain, *J. Geol. Soc. London.* **146,** 617-630 (1989).
15. Chumakov I.S, Yutsis V.V. Pliocene-Quaternary basins of the Eastern Mediterranean. *Bull. MOIP.otd.geol.,* **56,** 14-21 (russ.) (1981).
16. Doblas M., Lopez Ruiz J., Hoyos M., Martin C., and Cebria J.M. Late Cenozoic indentation/escape tectonics in the Eastern Betic Cordillera and its consequences on the Iberian forland, *Estudios geol.,* **47,** 193-205 (1991).
17. Eldholm O., Thiede J., Taylor E. Evolution of the V ring volcanic margin. Proceedings of the Ocean Drilling Program, *Scientific Results,* **104,** 1033-1065 (1989).
18. Eldholm O., Thiede J., Taylor E. The Norwegian continental margin: tectonic, volcanic and paleoenvironmental framework. Proceedings of the Ocean Drilling program, *Scientific Results,* **104,** 5-25 (1989).
19. Faleide J.l., Myhre A.M., Eldholm O. Early Tertiary volcanism at the western Barents Sea margin. Early Tertiary Volcanism and the Opening of the NE Atlantic, *Geological Society Special Publication,* **39,** 135-146 (1988).
20. Finetti I., Bricci G., A.Del Ben, Pipan M., and Xuan Z. Geophysical study on the Black Sea. Monograph on the Black Sea. *Boll.Geop.Teor.Appl., Trieste,* **XXX: 117-118,** 197-324 (1988).
21. Franke W. Tectonostratigraphic units in the Variscan units in the Variscan belt of Central Europe, *Geol. Soc. Am. Spec. Pap.* **230,** 67-90 (1989).
22. Frish W., Menot R.-P., Neubauer F., von Raumer J.F. Correlation and evolution of the Alpine basement *Schweiz. Mineral. and Petrogr. Mitt.* **Bd. 70. H. 2,** 265-285 (1990).
23. Gebauer D., Quadt A., Compston W., et al. Archaean Zircons in a retrograded "Caledonian" eclogite of Gottard Massif (Central Alps, Switzerland), *Schweiz. Mineral. and Petrogr. Mitt.* **Bd. 68**, 485-490 (1990.)
24. Gudlaugsson S.T., Gunnarsson K., Sand M., Skogseid J. Tectonic and volcanic events at the Jan Mayen Ridge microcontinent Early Tertiary Volcanism and the Opening of the NE Atlantic, *Geological Society Special Publication,* **39,** 85-93 (1988).
25. Guschenko O.l., Mastryukov A.O.,Petrov V.A. The structure of recent stress fields of seismic active zones in the eastern Mediterranean belt. *Doklady AN SSSR,* **312 (4P),** 830-835 (russ.) (1990).
26. Heinisch H. Hinweise auf die Existenzeines passiven Kontinentalrandes in Alt Palaozoikum der Nordlichen Grauwackenzone - Ostalpen. *Schweiz. Mineral and Petrogr. Mitt.* **Bd. 68,** 407-418 (1988).
27. Hippolyte J.-C., Angelier J., Bergerat F., Nury D., and Guieu G. Tectonic-stratigrafic record of paleostress time changes in the Oligocene basins of the Provence, southern France. *Tectonophysics,* **226,** 15-33 (1993).
28. *International Geological-Geophysical Atlas of the Atlantic Ocean.* Udintsev G.B. (ed.) IOC (of UNESCO), Min. Geol. USSR, AC. Sci. USSR, GUGK USSR, Moscow, 1989-1990. 158 p. (1990).
29. Jackson H.R., Faleide J.l., Eldholm O. Crustal Struture of the Sheared Southwestern Barents Sea Continental Margin. *Marine Geology,* **93,** 119-146 (1990).
30. Johansen B., Eldholm O., Talwani M. et al. Expanding spread profile at the nothern Jan Mayen Ridge. *Polar Research,* **6,** 95-104 (1988).
31. Jongsma D., Hinte J., Woodside J. Geological structure and neotectonics of the North African continental margin south of Sicily. Mar. and Petrol. *Geol.,* **2:2,** 156-179 (1985).
32. Kenyon N., Belderson R. *Young compressional structures of the Calabrian, Hellenic and Cyprus outer ridges.* Str.history of the Medditerranean basins. Paris, P.P. 233-240 (1977).
33. Kopp M.L. Megastructures of the horisontal extrusion in the Alpine-Himalaya orogenic belt. Autoref. of diss. *Thesis of the Dr.* Sci. Moscow, 35 p. (russ.) (1991).
34. Kopp M.L., Leonov Yu.G., Adjamian J. Deformation of the western Arabian plate as a result of strike-slip along the Levant fault. *Geotectonics,* **3,** 61-76 (russ.) (1994)
35. Korchemagin V.A., Emets V.S. Peculiarites of evolution of the tectonic structure and stress field of Donbass and eastern Priazovie. *Geotectonics,* **3,** 49-55 (russ.) (1987)
36. Leonov Yu.G. Stress fields in lithsphere and intraplate tectonics. *Geotectonics,* **6P,** 3-21 (russ.) (1995)

37. Le Pichon X., Angelier J. The Hellenic arc and trench system: a key to the neotectonic evolution of the Eastern Mediterranean area. *Tectonophysics*, **60:1/2,** 1-42 (1979)
38. Le Pichon X., Lyberis N., Angelier J., Renard V. Strain distribution over the East Mediterranean ridge: a synthesis incorporating new "Sea Beam" data, *Tectonophysics*, **86,** 243-274 (1982).
39. Letouzey J., Tremolieres P. Paleostress around the Mediterranean since the Mesozoic from microtectonics: comparison with plate tectonic data. *Rock. Mech.* **9:** 173-192 (1980).
40. Malod Y.A., Mauffret A. Iberian plate motions during the Mesozoic. *Tectonophysics*, **184:3/4,** 261-278 (1990).
41. Malovitsky Ya.P. *Tectonics of the Mediterranean bottom*. Moscow, Nedra, 96 p. (rus.) (1978)
42. Marques F.O., Ribeiro A., Munha J.M. Geodynamic evolution of the Continental Allochthonous Terrane (CAT) of the Braganca Nappe Complex, NE Portugal, *Tectonics*. **15:4,** 747-762 (1996).
43. Matte Ph. Tectonics and plate tectonic model for the Variscan belt of Europe. *Tectonophys.* **126,** 141-155 (1986).
44. Matte Ph. Accretionary history and crustal evolution of the Variscan belt in Western Europe, *Tectonophys.* **196,** 309-337 (1991).
45. Murillas Y. et al. Structure and evolution of the Galicia Interior Basin (Atlantic Western Iberian continental margin). *Tectonophysics*, **184:3/4,** 297-319 (1990).
46. Myhre A.M. Marine geophysical studies in the Norwegian-Greenland Sea and adjacent margins. *Doctorate thesis*, University of Oslo. 201 p (1984).
47. Neumann E.-R., Schilling J.-G. Petrology of basalts from Mohns-Knipovich Ridge; The Norwegian-Greenland Sea. *Contrib. Mineral. Petrol.*, **85,** 209-223 (1984)
48. Paquette J.L., Peucat J-J., Bernard-Griffiths J. et al. Evidence for old precambrian relicts shown by U-Pb zircon dating of eclogites and associated rocks in the Hercynian belt of South Brittany, France, *Chem. Geol. (Isotope Geosci. Section)*. **52:2,** 203-216 (1985).
49. Paquette J.L., Bale P., Ballevre M. et al. Geochronologie et geochemie des eclogites du Leon: Nord-Ouest du Massif Armorican, *Bull. Mineral.* **110,** 683-696 (1987).
50. Patchett P.J., Thorpe R.S., Beskinsale R.D. et al. Crustal growth and late Precambrian–early Paleozoic plate tectonic evolution of England and Walls, *J. Geol. Soc. London*. **141:Pt.3,** 521-536 (1984).
51. Perfiliev A.S., Akhmetiev M.A., Geptner A.R. et al. Miocene basalts of Iceland and problems of spreading. *Ac. Sci. USSR. Transactions*. **461,** 207 (russ.) (1991).
52. Peters J., Huson W. The Pliny and Strabo trenches (eastern Mediterranean): investigations of seismic reflection data and Sea Beam bathymetric maps. *Mar. Geol.*, **64:1/2,** 1-17 (1985).
53. Peucat J.J., Bernard-Griffiths., Gil Ibarguchi J.I. et al. Geochemical and geochronological cross-section of the deep Variscan crust: The Cabo Ortegal high-pressure nappe (NW Spain), *Tectonophys*. **177,** 263-292 (1990).
54. Pique A., Bossiere G., Bouillin J.-P. et al. Southern margin of Variscan belt: the north-western Gondwana mobile zone (Eastern Marocco and Northern Algeria), *Geol. Rund.* **Bd. 82,** 432-439 (1993).
55. Rastsvetajev L.M. The main features of the structural pattern of Earth's surface and its dynamic interpretation. Problems of global correlation of geologic fenomena. Trans. of Geol. Ins. of Academy of Sci. of the USSR, Moscow, Nauka, **340:** 145-194(rus.) (1980).
56. Rehault J., Moussat E., Fabri A. Structural evolution of the Tyrrhenian back-arc basin. *Mar. Geology,* **74:1/2,** 123-150 (1987).
57. Reuter C.-D., Eisbacher G. Pantelleric rift-crustal extension in a convergent intraplate setting. *Geol. Rdsch.*, **Bd.74: h.3,** 585-597 (1985).
58. Schilling J.-G., Zajak M., Evans R. et al. Petrology and geochemical variations along the mid-Atlantic ridge from 29 N to 73'N. *Am. J. Sci.*, **283,** 510-586 (1983).
59. Skogseid J., Eldholm O. Early Cainozoic evolution of the Norwegian volcanic passive margin and the formation of marginal highs. Early Tertiary volcanism and the opening of the NE Atlantic, *Geological Society Special Publication*, **39,** 49-56 (1988).
60. Spakman W., Wortel M., Vlaar N. The Hellenic subduction zone: a tomographic image and its geodynamic implications. *Geophys. Res. Lett.*, **15,** 60-63 (1988).
61. Spencer A.M., Home P.C., Berglund L.T. Tertiary structural developement of the western Barents Shelf, Troms to Svalbard. *Petroleum Geology of the North European Margin*. p.p. 199-209 (1984).
62. VaïG.B., Cocozza T. Tentative schematic zonation of the Hercinian chain in Italy *Bull. Soc. Geol. Fr. V111. T. 2.* **28,** 95-114 (1986).
63. White R.S., Spence G.D., Flower S.R. et al. Magmatism at rifted continental margins. *Nature*, **330:6147,** 439-444 (1987).
64. Zonenshain L.P., Le Pichon X. Deep basins of the Black Sea and Caspian Sea as remnants of Mesozoic back-arc basins. *Tectonophysics*, **123:1/4,** 181-240 (1986).

Proc. 30th Int'l Geol. Congr., Vol.7, pp. 157-167
Xu Zhiqin, Ren Yufeng and Qiu Xiaoping (Eds.)

Regional Geological Characteristics of China

MA LIFANG, MIN LONGRUI, DING XIAOZHONG
Geological Institute of CAGS, Baiwanzhuang Road, Beijing 100037, *CHINA*

Abstract

The 1:2,500,000 Geological Map of China is a large wall map which displays the geology of China. It is compiled by Ma Lifang (Institute of Geology, CAGS) and other seven colleagues of her such as Min Longrui, Ding Xiaozhong and so on. This map reflects comprehensively and systematically the recent achievement both in regional geological survey and in geological research during the past forty years, especially the late twenty years[1—27]. The date of data collected in it may be up to the end of 1996. In this map, it is emphasized particularly on its scientific and exact characteristics. Therefore, the map is compiled under the guidance of new academic viewpoints, and its complication is upon the bases of the careful analyses and the deepgoing studies on stratigraphy, igneous rocks and tectonics of different geological units. Besides, the map also expresses the sedimentary types of some strata, rocks types of some igneous rocks, metamorphic facies groups, genetic types of the Quaternary and its isopachous line as well as some typical drill holes, etc. For the representation of the legend, a combination form, which combines together those of stratigraphy, igneous rocks, structural deformation and metamorphism, is adopted in order to reveal the evolutional processes of the different tectonic units of three dimensions more comprehensively and visually. Some geological information concerning enviromental geology, catastorphic geology and global changes are added. The geological information since Quaternary are emphasized in order to reflect the formative processes and the external dynamical conditions of the Quaternary geological bodies.

Keywords: Regional Geology, Precambrian, Jurassic to Cretaceous, Quaternary, Tectonic Division, Platform, Fold Belt

The Chinese continent developed and evolved gradually through a long-term and mutual interaction of the Siberia, North China, Tarim, Yangtze, South China, India and the Pacific plates. Based on the overall characteristics of the space—time development in sedimentary association, igneous activities, metamorphism and tectonic movement, the Chinese continent may be divided roughly into two classes, i. e. the platform and the fold areas. The former includes the North China platform, the Tarim platform and the Yangtze platform area, all of which constitute the main part of the Chinese continent. The latter includes the Junggar—Inner Mongolia—Hinggan fold belt, the Kunlun—Qingling fold belt, the Qinghai—Xizang—West Yunnan

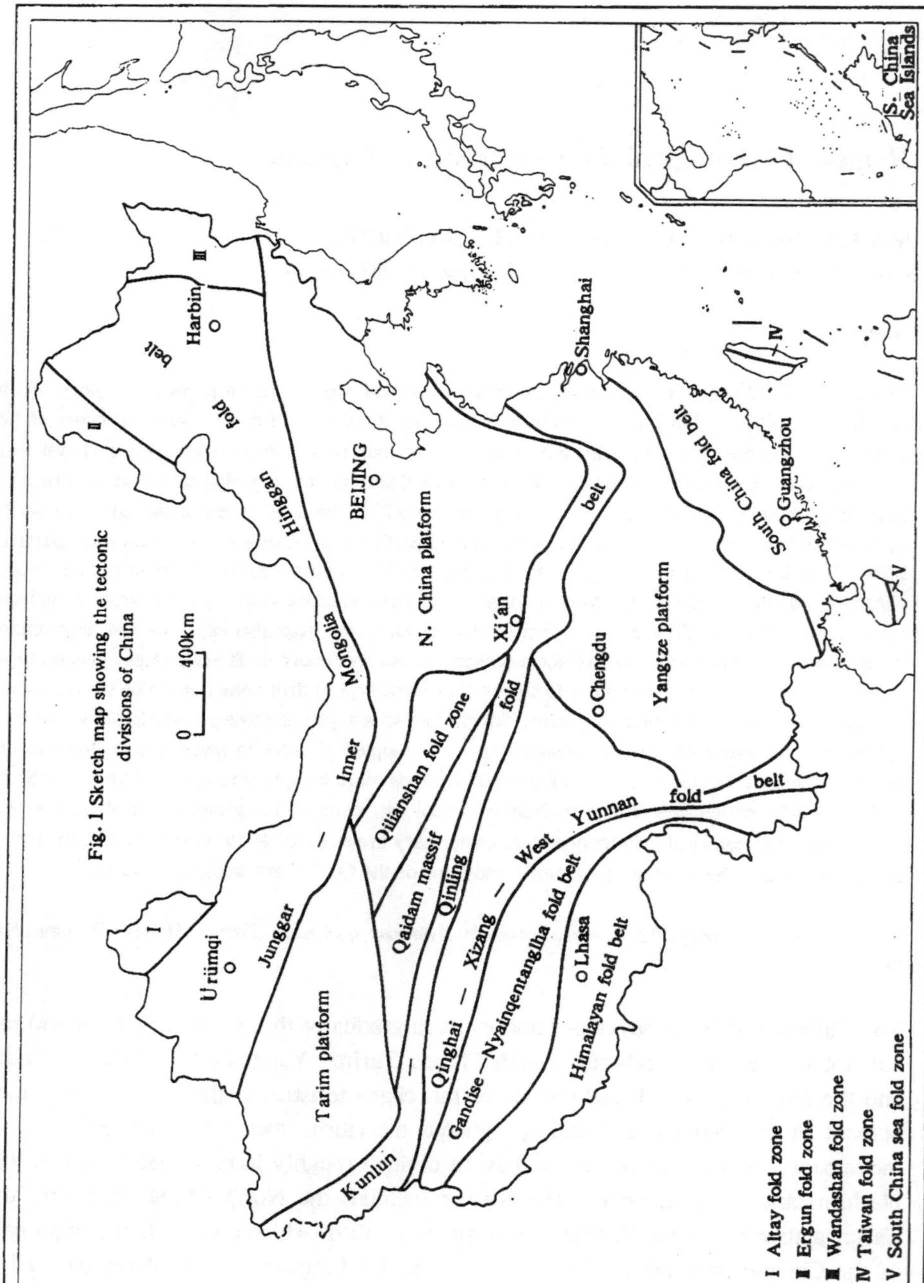

Fig. 1 Sketch map showing the tectonic divisions of China

fold belt, the Gandise—Himalaya fold belt, the South China fold belt, the Wandashan fold zone and the Taiwan fold zone[30,31]. (Fig. 1)

The division and correlation of the Precambrian have been the focal points of research at home. The successive discoveries of the oldest surface crustal rocks (Caozhuang Group) in East Hebei and the granitic oldland crustal relics near Anshan in recent years allow us to make a three-fold division for the Archean Eonothem based on 3500 Ma and 3000 Ma respectively. The Proterozoic and Archean Eonothems are still divided by 2600 Ma. All of these age data only represent roughly the division age (Table 1). The division between the Jurassic and Cretaceous systems is still a problem that has been controverted for a long period. Recently, along with the progress in research of fossil organism and the improvement in measurement of isotopic age, the period of the Rehe fauna-bearing strata in the N, NE, and SE China parts become clearer and clearer. Considering there is no unified source and understanding on the data, it is treated as J_3-K_1 in this geologic map (Table 2). The Quaternary is the youngest unit in geological history. The human beings are active in this time. In recent years, with the studies of global change, environment geology, upheaval of Qinghai—Xizang (Tibet) plateau and ancient humen beings, it is payed more attention to study the Quaternary[28]. Quaternary developed very well in China, it covers about quart of the land. In this map, the Quaternary is subdivided into series and different genetic types are showed. 2. 48 Ma is taken as the bottom boundary of the Quaternary, whose main basis are as follows: 1) The fossils such as Equus Sammeniensis, Hyalinea ballica in the Nihewan formation of North China and in the Yuanmo formation of Yunnan are about 2. 4 Ma. ;2) The bottom boundary of loess—paleosol in China is 2. 48 Ma. The loess is the only continental accumulation which can be compared to deep sea sediments in oxygen isotopes contains plentiful climate information (Fig. 2).

1. North China platform: The basically consolidated stable massif, formed after the Luliang movement, constitutes the main part of North China plate. Its Archaean outcropping area is the most complete and the best developed one. In the southern Inner-Mongolia, from Sonid Yonqi to the south of the Xar Moron River is the northern margin of the North China plate, and there is mainly the Caledonian fold zone.

2. Tarim platform: It was consolidated in the Jinning movement (850 Ma). Since Tertiary, along with the large amplitude uplifting of the Qingzang plateau, the Tarim subsided relatively to form the largest inland basin in China. The buried depth of its basement is 8—10 Km and it is uplifted in the west but shows as a superimposed fault depression in the east[32].

3. Yangtze platform: It is separated from the Qinling fold belt by the Shanyang—Tongcheng fault, and it is demarcated from the Qingzang—W. Yunnen fold belt by

Table 1 The division and correlation of Precambrian in China

Age			Tianshan—Xinggan: Tianshan	Tianshan—Xinggan: Jiamusi	North China: E. Hebei	North China: Wutai—Taihang Mts.	North China: E. Shandong	North China: W. Henan	North China: S. Jilin	North China: Yinshan Mts.	Tarim: Kuluktag	Kunlun—Qinling: C. Qilian	Kunlun—Qinling: Qinling	Kunlun—Qinling: Dabie Mts.	Yangtze: W. Hubei	Yangtze: Hunan-Guizhou-Guangxi	Yangtze: Jiangsu-Anhui
Proterozoic	Late	Sinian Late	Kailaketi Gr.	Zhangguangcailing Gr.				Luoquan Fm.	Qinggouzi Fm.		Kuluktag Gr.	Duoruonuoer Gr.	Taowan Gr.	Baizhaoshan Fm.; Chahe Fm.	Dengying Fm.; Doushantuo Fm.	Liuchapo Fm.; Doushantuo Fm.	Piyuanchun Fm.; Lantian Fm.
		Sinian Early					Shiwangzhuang Fm.; Fulaishan Fm.; Tongjia Fm.	Dongjia Fm.; Huanglianduo Fm.	Badaojiang Fm.; Wanlong Fm.; Qiaotou Fm.	Sernag Gr.				Hongan Gr.	Nantuo Fm.; Liantuo Fm.	Hongjiang Fm.; Jiangkou Fm.	Sujiawan Fm.; Zhougang Fm.
		Qingbaikou stage	Kushitai Gr.	Majiajie Gr.	Qingbaikou Gr.		Erqingshan Fm.; Heishangong Fm.	Luoyukou Gr.	Nanfen Fm.; Diaoyutai Fm.; Baifangzi Fm.		Pargangtag Gr.	Gongcha Gr.			Macaoyuan Fm.	Banxi Gr.	Zhangbaling Gr.
	Middle	Jixian stage	Koksu Gr.	Heilongjiang Gr.	Jixian Gr.		Penglai Gr.	Ruyang Gr.	Seluohe Gr.	Zhartai Gr.	Arjigan Gr.	Tuolainanshan Gr.	Kuanping Gr.		Shennongjia Gr.	Lengjiaxi Gr.	
		Changcheng stage	Tekes Gr.		Changcheng Gr.			Xungger Gr.			Yangjibulak Gr.	Danghe Gr.					
	Early	Hutuo stage	Nalati Gr.	Xingdong Gr.	Zhuzhangzi Gr.	Hutuo Gr.	Fenzishan Gr.	Songshan Gr.	Laoling Gr.	Hongqiyingzi Gr.	Xingditag Gr.	Yiemananshan Gr.	Qinling Gr.	Dabie Gr.	Shuiyuesi Gr.		Donghai Gr.
		Wutai stage			Shuangshanzi Gr.	Wutai Gr.			Ji'an Gr.								
Archean	Late			Mashan Gr.	Dantazi Gr.	Fuping Gr.	Yishui Gr.	Dengfeng Gr.	Jiapigou Gr.	Wulashan Gr.	Daglagbulak Gr.				Dongconghe Fm.		
	Middle				? Qianxi Gr.				? Longgang Gr.	Jinning Gr.							
	Early				? Chaozhuang Gr.												

Table 2 The division and correlation of Jurassic to Cretaceous in Eastern China

Strata		North Hebei	West Liaoning	South Jilin	Da Hinggan Ling		South Jiangsu	Shandong	West Zhejiang	Fujian	Jiangxi	Guangdong
					South	North						
Cretaceous	Upper	Tujingzi Fm. Ximalin Fm.		Sankeyushu Fm.		Qingyuan-gang Fm.	Chishan Fm. Pukou Fm.	Wangshi Fm.	Qujiang Gr. Yongkang Gr.	Shimaoshan Gr.	Nanxiong Fm. Ganzhou Fm.	Nanxiong Fm.
	Lower	Qingshila Fm. Nandian Fm.	Sunjiawan Fm. Fuxin Fm. Jiufutang Fm.	Heiwaizi Fm. Hengtong-shan Fm.		Yimin Fm. Damoguaihe Fm.	Niangniang-shan Fm. Gushan Fm. Dowangshan Fm. Yunheshan Fm.	Qingshan Fm. Laiyang Fm.	Hengshang Fm. Shouchang Fm.	Bantou Fm.	Huobashan Gr. Ehuling Fm.	Guancaohu Gr.
	U. Jurassic—L. Cretaceous	Huajiying Fm. Xiguayuan Fm. Dabeigou Fm.	Jianchang Fm. Jingangshan Fm. Yixian Fm.	Xiahuapi-dianzi Fm. Baodaqiao Fm. Dashatan Fm. Lamenzi Fm.	Meiltu Fm. Bayanggol Fm.	Yilikede Fm. Shangkuli Fm. Murui Fm.	Longwang-shan Fm. Xihengshan Fm.	Xiwa Fm. Mongyin Fm.	Huangjian Fm. Laochun Fm.	Nanyuan Fm. Changlin Fm.	Daguding Fm. Huajiancao Fm. Zhoujiayuan Fm.	Gaojiping Gr.
Jurassic	Upper	Zhangjiakou Fm.			Manit Fm. Manketouobo Fm.	Tamulangou Fm.		Fenshuling Fm.				
	Middle	Houcheng Fm. Tiaojishan Fm. Jiulongshan Fm. Longmen Fm.	Tuchengzi Fm. Lanqi Fm. Haifanggou Fm.	Houjiatun Fm. Xiajiajie Fm. Wangjialou Fm.	Xinmin Fm. Wanbao Fm.	Nanping Fm. Taipingchuan Fm.	Xiangshan Gr.	Santai Fm. Fangzi Fm.	Yujianshan Fm.	Zhangping Fm.	Luoao Fm.	Maxinping Gr.
	Lower	Xiahuayuan Fm. Nandaling Fm.	Beipiao Fm. Xinglonggou Fm.	Yihe Fm.	Hongqi Fm.			Wennan Fm.	Majian Fm.	Lishan Fm.	Linshan Fm.	Jinji Gr.

Fig. 2 Sketch map for the climatic evolution of the loess—paleosol series of China

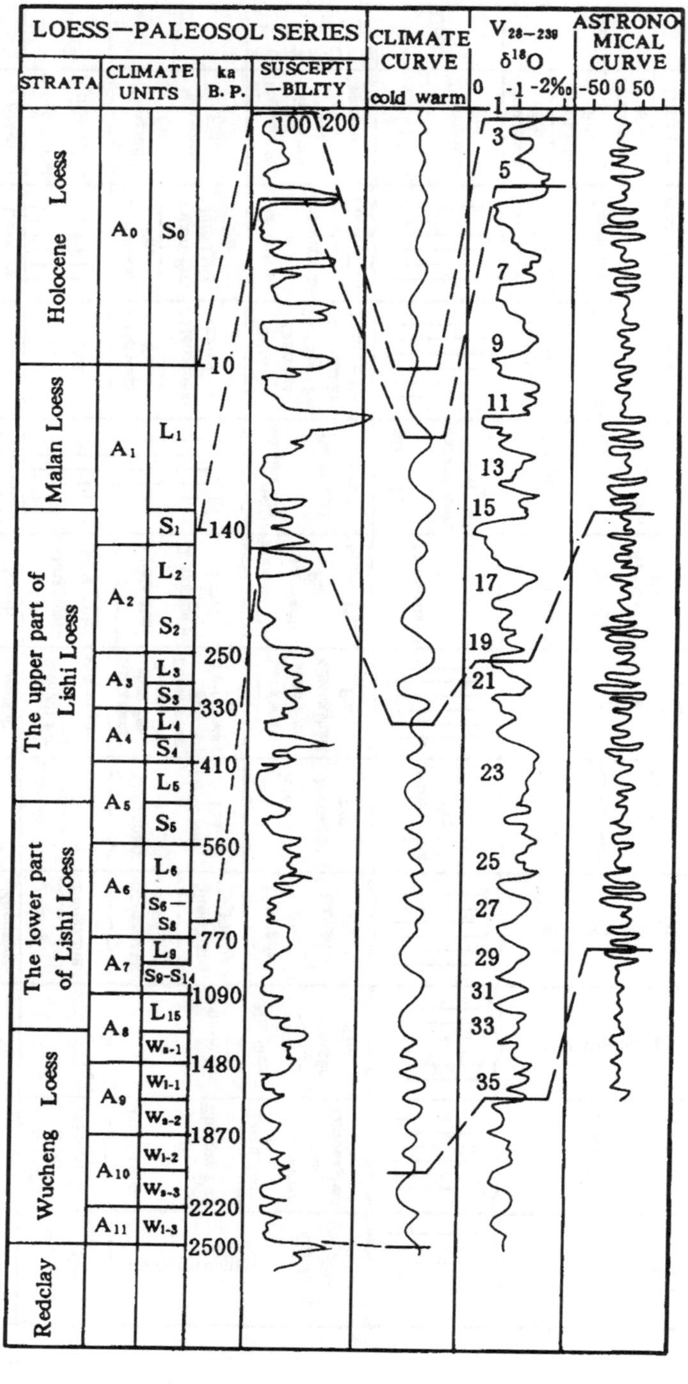

the Longmenshan—Honghe fault zone in the west; in the southeast, it is adjacent to the South China Caledonian fold system by the Shaoxing—Jiangshan fault. This area was formed after the Jinning movement, but according to the 2957 Ma value obtained from the Kangding Gr. in southern Sichuan and a lot of 1700 Ma value obtained in other places, clearly this indicates that some places belonged to the stable area consolidated in the Luliang movement. The Sinian system—middle Triassic series are typically cover deposites, among which the Sinian—Cambrian section in Yangtze Gorges is one of the stratotype sections.

The Qaidum massif in the southwest might be split away from the southwest margin of the North China plate in Late Proterozoic era. The Qilianshan Caledonian fold zone was a sea trough formed then. It was folded and uplifted in Late Silurian. The occurrence of mollasse formation in Middle Devonian indicates that the Qaidum massif and the North China platform were united to form a unified continental crustal area.

4. Junggar—Inner Mongolia—Hinggan fold belt: It is a part of the Paleo-Asia continental margin´ s accretionary fold belt among the Siberia, Tarim and North China plates, among which, there scatter the Junggar, Xilinhot and Jiamusi minor massifs. The accretionary evolution of the continental margin occurred in the Caledonian and Early Variscan stages. In Early Carboniferous, among the Siberia plate, Tarim plate and North China plate there took place collision and convergence, thus leading to the development of folding, and faulting, also there occurred intensive volcanic activities. In the meantime, there were many complicated and various types of metamorphism, which constituted the main Plaeozoic tectonmagmatic belt in China. After the Variscan stage, in the west segment, under the combined compression of the Siberia, Kazakhstan, and India plates there formed the alternating range and basin tectonic pattern, along with a series of thrust-nappe structures and large scale strike-slip faults; in the eastern segment, besides the influence coming from the Siberian plate, there underwent compressions for many times from the Pacific plate on the SE which produced the mutual composite tectonic pattern among the EW-, NE- and NNE- ward tectonics.

5. Kunlun—Qinling fold belt: It was a crustal subduction zone located between the Tarim and the North China as well as the Yangtze plates. Therefore, the internal composition and structure of this belt are very complicated, and leaving many geologic problems to be further explored. Based on the data available, we can see that it is a composite orogenic belt resulted from orogenic movements taking place in the Jinning, Caledonian, Variscan and Indosinian stages. It is composed mainly of a series of Palaeozoic fold zones, thrust-fractures, strike-slip faults and ophiolite zones; and there are intruded in them large amounts of Palaeozoic acidic, basic and ultrabasic intrusive rocks. Being truncated by the Tancheng—Lujiang fault, the east segment

and is displaced as far as southeastern Shandong, then it strikes NEE; the west segment is truncated by the Altun fault. The Kunlun fold belt may be divided into the southern and the northern parts by the Kangxiwa—Middle Kunlun fault, and the northern Kunlun is a Variscan fold belt, while the southern Kunlun is a Variscan, Indosinian fold belt. The Qinling fold belt lay somewhere between the North China and Yangtze plates. The existing rocks are generally composed of the two great plate margins in the south and north, while the southern and northern Qinling are divided by the Shangcheng—Danyang fault. The northern Qinling is a Caledonian orogenic belt, while southern Qinling a Variscan, Indosinian fold belt. Its basement is composed of the Upper Archean and Lower Proterozoic Dabieshan Gr. and Jiaonan Gr. as well as the Middle Proterozoic Wudang Gr. , the recent studies prove that there exists a large ductile decollement shear zone between the basement and the cover. Meantime, there were massive Indosinian granitoid infrusions. Being affected by the NNW movement of the Paleo-Pacific plate, the Qinling—Dabieshan zone continued during the post Late Trassic to produce thrusts, decollement and overthrusts. It has been inferred lately that younger strata might have existed beneath the Dabieshan Group.

6. Qinghai—Xizang—West Yunnan fold belt: It is separated from the Kunlun fold belt by the Xugui—Maqen fault in the north, but connected with the Gandise—Himalayan fold area by the Bangongco—Nujiang fault. This belt is composed of the Bayan Har fold zone and the Tanggula fold zone, as well as some intermediate massifs, nappe structures, ophiolite, melange and tectonomagmatic belts. The two fold zones are demarcated by the Hohxil—Jinshajiang fault. The Bayan Har fold zone belonged originally to the west margin of the Yangtze plate, but it formed the Indosinian fold zone by splitting apart from the Yangtze continent at the beginning of Late Palaeozoic. The Tanggula fold zone is composed mainly of Upper Triassic and Jurassic fold zones, thrust faults and ophiolite belts, and in it, there injected a series of granitoid rock masses. Under the huge thick cover, there might exist the Precambrian basement. During Late Triassic, the Jinshajiang block subducted southward and closed up, when the Tanggula fold zone coalesced together with the Bayan Har fold zone.

7. Gandise—Himalayan fold belt: It is a split portion separated from the northern margin of the Gondwanaland. It may be divided into the Gandise—Nyainqentanglha fold belt and the Himalayan fold zone by the Yarlung zangbo Jiang belt. The Nyainqentanglha fold zone is a Late Yanshanian fold zone. In Late Palaeozoic, there occurred the glaciomarine deposits and cold water fauna with the characteristics of Gondwanaland. In Jurassic, the trench—arc—basin system made its appearance, and deposited the hugely thick turbidites, containing great quantities of basic—ultrabasic rocks, radiolarian siliceous rocks and melange masses. The famous Gandise volcanomagmatic arc was formed in the Late Yanshanian and Early Himalayan stages.

The Himalayan fold zone is a Cenozoic one and it was connected with the Indian plate by the main boundary fault. On the Precambrian crystal basement, there deposited the platform cover of a Palaeozoic carbonate rocks intercalated with clastic rocks. By the end of Permian and the beginning of Triassic, along with the opening of the Yarlung zangbo Jiang—Tethys domain, there in the Yarlung zangbo Jiang area developed the active type flysch deposits, together with volcanic and allochthonous rocks. In Late Cretaceous, the Indian plate drifted northward, the Tethys closed gradually, and the ophiolite belt appeared. In Eocene, the India plate and Eurasia plate converged, the Tethys ocean disappered, Qinghai—Xizang plateau upheavaled. In Plicene—Early Pleistocene, the Himalayas rise up rapidly. In Middle—Late Pleistocene, the Himalayas was upheavaling contineously, at last, the Qomolangma peak becomes the highest peak in the world.

8. South China fold belt: Its main part belonged to the Caledonian fold belt, but under the intensive influence of the Variscan, Indosinian and especially the Yanshanian tectonomagmatism, there presented the polystage, mutually superimposed, composite tectonic pattern. During Late Triassic to Late Tertiary, under the influence of the Pacific plate, there formed a series of fault basins trending NE or NNE, accompanied with intensive tectonic movement and magmatic activities.

9. Wandashan fold belt: It is a part of Xihuotalin fold belt, and was a continental margin accretion belt formed along the eastern periphery of the Asia continent during Late Jurassic—Early Cretaceous. It is composed mainly of Carboniferous—Permian limestone and greenschist, Middle — Upper Triassic radiolarian — bearing siliceous rock, turbidite and Lower — Middle Jurassic clastic and volcanic rocks. Some of these rock formations appeared as allochonous rock masses in Late Jurassic strata. In this area, the thrust and nappe structures are very complicated, and there were also granitoid emplacement taking place in the Indosinian and Yanshan stages.

10. Taiwan fold zone: It is a component part of the West Pacific island arc fold belt, and divided by the Taidong great longitudinal rift belt into the Central mountain fold zone and the Coastal Range fold zone in Eastern Taiwan. The latter connected with the Phillipine Luzon island arc and belonged to the Phillipine plate; the former, the Central mountain, connected with the Taoyu island upwarp in the north belonged to the Eurasian plate, while the great longitudinal rift belt was a convergent belt of the Phillipine plate and the Eurasian plate. The Taidong Coastal Range fold belt is composed of Late Tertiary clastics, and island—arc volcanics; divided into the east and west parts: the former made up chiefly of Miocene Qimi volcanics and Pliocene —Pleistocene flysches. The Pliocene ophiolite melange belt in the SE side is made up of oceanic crust substance which was brought there in by the collision and subduction of the Phillipines plate[29].

CHIEF REFERENCES

1. Bureau of Geology and Mineral Resources of Jiangsu Province, *Regional Geology of Jiangsu Province and Shanghai Municipality*, Geological Publishing House(1984).
2. Bureau of Geology and Mineral Resources of Jiangxi Province, *Regional Geology of Jiangxi Province*, Geological Publishing House(1984).
3. Bureau of Geology and Mineral Resources of Guangxi Zhuang Autonomous Region, *Regional Geology of Guangxi Zhuang Autonomous Region*, Geological Publishing House(1985).
4. Bureau of Geology and Mineral Resources of Fujian Province, *Regional Geology of Fujian Province*, Geological Publishing House(1985).
5. Bureau of Geology and Mineral Resources of Anhui Province, *Regional Geology of Anhui Province*, Geological Publishing House(1987).
6. Bureau of Geology and Mineral Resources of Guizhou Province, *Regional Geology of Guizhou Province*, Geological Publishing House(1987).
7. Bureau of Geology and Mineral Resources of Hunan Province, *Regional Geology of Hunan Province*, Geological Publishing House(1988).
8. Bureau of Geology and Mineral Resources of Guangdong Province, *Regional Geology of Guangdong Province*, Geological Publishing House(1988).
9. Bureau of Geology and Mineral Resources of Jilin Province, *Regional Geology of Jilin Province*, Geological Publishing House(1988).
10. Bureau of Geology and Mineral Resources of Zhejiang Province, *Regional Geology of Zhejiang Province*, Geological Publishing House(1989).
11. Bureau of Geology and Mineral Resources of Shaanxi Province, *Regional Geology of Shaanxi Province*, Geological Publishing House(1989).
12. Bureau of Geology and Mineral Resources of Liaoning Province, *Regional Geology of Liaoning Province*, Geological Publishing House(1989).
13. Bureau of Geology and Mineral Resources of Henan Province, *Regional Geology of Henan Province*, Geological Publishing House(1989).
14. Bureau of Geology and Mineral Resources of Shanxi Province, *Regional Geology of Shanxi Province*, Geological Publishing House(1989).
15. Bureau of Geology and Mineral Resources of Gansu Province, *Regional Geology of Gansu Province*, Geological Publishing House(1989).
16. Bureau of Geology and Mineral Resources of Hubei Province, *Regional Geology of Hubei Province*, Geological Publishing House(1990).
17. Bureau of Geology and Mineral Resources of Yunnan Province, *Regional Geology of Yunnan Province*, Geological Publishing House(1990).
18. Bureau of Geology and Mineral Resources of Hebei Province, *Regional Geology of Hebei Province, Beijing Municipality and Tianjing Municipality*, Geological Publishing House(1990).
19. Bureau of Geology and Mineral Resources of Ningxia Hui Autonomous Region, *Regional Geology of Ningxia Hui Autonomous Region*, Geological Publishing House(1990).
20. Bureau of Geology and Mineral Resources of Inner—Mongulia Autonomous Region , *Regional Geology of Inner—Mongulia Autonomous Region*, Geological Publishing House(1991).
21. Bureau of Geology and Mineral Resources of Sichuan Province, *Regional Geology of Sichuan Province*, Geological Publishing House(1991).
22. Bureau of Geology and Mineral Resources of Qinghai Province, *Regional Geology of Qinghai Province*, Geological Publishing House(1991).
23. Bureau of Geology and Mineral Resources of Shandong Province, *Regional Geology of Shandong*

Province, Geological Publishing House(1991).
24. Bureau of Geology and Mineral Resources of Heilongjiang Province, *Regional Geology of Heilongjiang Province*, Geological Publishing House(1993).
25. Bureau of Geology and Mineral Resources of Beijing Municipality, *Regional Geology of Beijing Municipality*, Geological Publishing House(1991).
26. Bureau of Geology and Mineral Resources of Xizang (Tibet) Autonomous Region, *Regional Geology of Xizang (Tibet) Autonomous Region*, Geological Publishing House(1993).
27. Bureau of Geology and Mineral Resources of Xinjiang Uygur Autonomous Region, *Regional Geology of Xinjiang Uygur Autonomous Region*, Geological Publishing House(1993).
28. Zhang Zhongfu, *Geological Map of Quaternary of the People′s Republic of China and Adjacent Sea Area*, Cartographic Publishing House(1990).
29. He Chunxun , *Geological Introduction of Taiwan*, Published by Geological Survey of Taiwan (1986).
30. Cheng Yuqi et al, *Stratigraphic Law of China*, Geological Publishing House(1996).
31. Ma Lifang et al, *Geological Atlas of China*, Geological Publishing House(1997).
32. Chen Zhefu, Wu Naiyuan et al, *Geological Map of Xinjiang Uygur Autonomous Region, China*, Geological Publishing House(1985).